T0255772

From Concepts to Code

The breadth of problems that can be solved with data science is astonishing, and this book provides the required tools and skills to a broad audience. The reader takes a journey into the forms, uses, and abuses of data and models, and learns how to critically examine each step. Python coding and data analysis skills are built from the ground up, with no prior coding experience assumed. The necessary background in computer science, mathematics, and statistics is provided in an approachable manner.

Each step of the machine learning lifecycle is discussed, from business objective planning to monitoring a model in production. This end-to-end approach supplies the broad view necessary to sidestep many of the pitfalls that can sink a data science project. Detailed examples are provided from a wide range of applications and fields, from fraud detection in banking to breast cancer classification in healthcare. The reader will learn the techniques to accomplish tasks that include predicting outcomes, explaining observations, and detecting patterns. Improper use of data and models can introduce unwanted effects and dangers to society. A chapter on model risk provides a framework for comprehensively challenging a model and mitigating weaknesses. When data is collected, stored, and used, it may misrepresent reality and introduce bias. Strategies for addressing bias are discussed. *From Concepts to Code: Introduction to Data Science* leverages content developed by the author for a full-year data science course suitable for advanced high school or early undergraduate students. This course is freely available and it includes weekly lesson plans.

Adam P. Tashman has been working in data science for more than 20 years. He is an associate professor of data science at the University of Virginia School of Data Science. He is currently Director of the Capstone program, and he was formerly Director of the Online Master's of Data Science program. He was the School of Data Science Capital One Fellow for the 2023–2024 academic year. Dr. Tashman won multiple awards from Amazon Web Services, where he advised education and government technology companies on best practices in machine learning and artificial intelligence. He lives in Charlottesville, VA with his wonderful wife Elle and daughter Callie.

From Concepts to Code
Introduction to Data Science

Adam P. Tashman

CRC Press
Taylor & Francis Group
Boca Raton London New York

CRC Press is an imprint of the
Taylor & Francis Group, an **informa** business

A CHAPMAN & HALL BOOK

Designed cover image: © Adam P. Tashman

First edition published 2024
by CRC Press
2385 NW Executive Center Drive, Suite 320, Boca Raton FL 33431

and by CRC Press
4 Park Square, Milton Park, Abingdon, Oxon, OX14 4RN

CRC Press is an imprint of Taylor & Francis Group, LLC

© 2024 Adam P. Tashman

Reasonable efforts have been made to publish reliable data and information, but the author and publisher cannot assume responsibility for the validity of all materials or the consequences of their use. The authors and publishers have attempted to trace the copyright holders of all material reproduced in this publication and apologize to copyright holders if permission to publish in this form has not been obtained. If any copyright material has not been acknowledged please write and let us know so we may rectify in any future reprint.

Except as permitted under U.S. Copyright Law, no part of this book may be reprinted, reproduced, transmitted, or utilized in any form by any electronic, mechanical, or other means, now known or hereafter invented, including photocopying, microfilming, and recording, or in any information storage or retrieval system, without written permission from the publishers.

For permission to photocopy or use material electronically from this work, access www.copyright.com or contact the Copyright Clearance Center, Inc. (CCC), 222 Rosewood Drive, Danvers, MA 01923, 978-750-8400. For works that are not available on CCC please contact mpkbookspermissions@tandf.co.uk

Trademark notice: Product or corporate names may be trademarks or registered trademarks and are used only for identification and explanation without intent to infringe.

Library of Congress Cataloging-in-Publication Data

Names: Tashman, Adam P., author.
Title: From concepts to code : introduction to data science / Adam P. Tashman.
Description: First edition. | Boca Raton FL : CRC Press, 2024. | Includes bibliographical references and index.
Identifiers: LCCN 2023049717 (print) | LCCN 2023049718 (ebook) | ISBN 9781032517957 (hardback) | ISBN 9781032517988 (paperback) | ISBN 9781003403982 (ebook)
Subjects: LCSH: Data mining. | Big data. | Electronic data processing.
Classification: LCC QA76.9.D343 .T3635 2024 (print) | LCC QA76.9.D343 (ebook) | DDC 005.7--dc23/eng/20240220
LC record available at https://lccn.loc.gov/2023049717
LC ebook record available at https://lccn.loc.gov/2023049718

ISBN: 978-1-032-51795-7 (hbk)
ISBN: 978-1-032-51798-8 (pbk)
ISBN: 978-1-003-40398-2 (ebk)

DOI: 10.1201/9781003403982

Typeset in CMR10
by KnowledgeWorks Global Ltd.

Publisher's note: This book has been prepared from camera-ready copy provided by the authors.

Access the Support Material: https://github.com/PredictioNN/intro_data_science_course/

To my father, Steven M. Tashman, who launched my quest by asking:

"What are you going to do with math?"

And to my mother, Linda E. Tashman, who encourages me to keep going.

Contents

Acknowledgments

I thank my wife, Elle Tashman, for reviewing the early chapters and providing support and encouragement. I am very grateful to Philip D Waggoner and J Gavin Wu for their thoughtful feedback on the manuscript. Efrain Olivares provided helpful, detailed ideas and feedback on reproducible data science. The School of Data Science at the University of Virginia provided an extremely supportive and enriching environment for this work. In particular, I thank Phil Bourne, Jeffrey Blume, Don Brown, Raf Alvarado, Brian Wright, Jon Kropko, Pete Alonzi, Siri Russell, and Emma Candelier. Boris Deychman and Judy Pann provided invaluable insight into model risk, and Greg van Inwegen generously shared knowledge on modeling and quantitative finance. I am very grateful to Lara Spieker for encouraging me to write this book and for shepherding it through the process. I thank the supporting team at Chapman and Hall/CRC. Finally, I wish to acknowledge the impact that Devin Chandler, Lavel Davis Jr., and D'Sean Perry had on this book. Following their heartbreaking tragedies, I felt compelled to write this book as a response. I send love to their families.

Preface

The amount of data generated in the world and the number of decisions to be made from this data has never been greater. Over 2.5 quintillion bytes of data are produced by humans every day [1], and new units of measure are on the way to quantify data of the future. From healthcare and finance to retail and marketing, every field that isn't in the technology sector is directly using technology, or likely should be using it. Many of these fields are spawning new fields with a suffix: finance has FinTech, insurance has InsurTech, and healthcare has HealthTech. There are even more specialized fields, like WealthTech, MarTech (marketing), and AdTech (advertising).

This is exciting, but it is also exceedingly difficult to find people who know the underlying field and the technology really well. In the early days of data science, there were some rare unicorns with strong quantitative and programming skills who could do wizardry in their industries. Perhaps the IT department lent a hand, or they were applied mathematicians or statisticians. One-off projects proved that understanding the organization's data could work wonders. Simple analytics, or perhaps some regression analysis, caught the attention of management. As the desire to repeat and productize the data assets grew, however, it became clear this was a full-time job, or perhaps it required an entire department of people. Data became a strategic asset, and the people needed to tap into this asset were in high demand.

As of 2024, demand for this kind of work and the people who bring the magic has never been greater. The job functions of the AI-driven world are better understood, and they have names like data scientist, data engineer, and machine learning engineer. These lucky people are some of the most desired knowledge workers in the world. Yet there is a massive shortage of them.

From the supply side, we still graduate generations of students doing math without computers. However, there is good news: as early as kindergarten, children are counting how many siblings they have, representing those counts with little tiles of paper, aggregating the tiles, and organizing them into columns to make a picture. What I'm saying is this: we have children all over the world taking their first steps in data science. They just need more resources and guidance.

As a passion project, I created a high school data science course with Matt Dakolios, who was one of my students from the University of Virginia (UVA) School of Data Science. As you might guess, it's a tremendous amount of work to create a high school course. Unfortunately for me, I can't seem to pick passion projects that are easy. The best I can hope for is that the projects don't flop. Eagerly following the students piloting the course, Matt and I found that it sparked an interest that didn't exist before. Some of them wanted to pursue data science in college, while others had a peaked interest in science or math.

Given this positive feedback, I have been working to get adoption of the course at other schools. The great part about "selling" a product is that the tough questions make you think and reflect (although it's best to do this ahead of time!). During one meeting with the director of a highly regarded, diverse school, he asked this direct question:

"We have 59 initiatives at my school currently. Why should we consider this one?"

It's a fair question, I thought, especially as we're reeling from a multi-year pandemic. I think it comes down to this, I said[1]:

"If you could engage a wide range of students, equip them with a skill set to make a substantial dent in the world's major problems, allow them to do something personally meaningful, present a career path that earns a nice living, all while applying and developing their skills in science and mathematics, would you be interested? Because that's data science."

This book aims to bridge the gap between a workplace in need of more data scientists, and a workforce that stands to greatly benefit in this field. However, it's bigger than this: even if you have no interest in ever working in data science, this book will expand your data literacy, and your understanding of how things work. I hope you enjoy this journey into the world of data.

---

[1]Well, I more or less said some of this, and the Dean of the UVA School of Data Science, Phil Bourne, jumped in with an assist. It was a team effort!

Symbols

Symbol Description

Set Theory

$\mathbb{Z}$	The set of integers
$\mathbb{R}$	The set of real numbers
$\mathbb{R}^+$	The set of positive real numbers
$[a, b]$	The real interval including a and b
(a, b)	The real interval excluding a and b
$\{\}$ or $\emptyset$	The empty set
$A \cap B$	Intersection of sets A and B
$A \cup B$	Union of sets A and B
A^c	Complement of set A
$A - B$	Elements in set A but not B
$|A|$	The number of elements in set A
$Y \subseteq X$	Y is a subset of X
$Y \subset X$	Y is a proper subset of X

Operations

$\sum_{i=1}^n$	Sum of terms from 1 to n
$\prod_{i=1}^n$	Product of terms from 1 to n

Functions and Metrics

$f(x)$	A function evaluated at x
f^{-1}	Inverse of the function f
$f \circ g$	Composition of the functions f and g
$ln(x)$	Natural logarithm of x
$\sigma(x)$	Sigmoid function
$d(x, y)$	Distance between x and y

Calculus

lim	The limit of a sequence or function
$f'(x)$	The function derivative
$\frac{dy}{dx}$	Derivative of y with respect to x
$\frac{\partial y}{\partial x}$	Partial derivative of y with respect to x
∇y	Gradient of y

Probability

Ω	The sample space (all possible outcomes)
$\mathcal{A}$	The set of all events
$P(A)$	The probability of event A
$P(A|B)$	The probability of event A conditional on B
$P(X = x)$	The probability that random vari-

able X takes value x

$H(\hat{p}_i, y_i)$ — Cross entropy loss

pmf — The probability mass function

pdf — The probability density function

cdf — The cumulative distribution function

$\binom{n}{y}$ — Number of ways to select y elements from n elements

$Bin(n, p)$ — Binomial distribution

$U(a, b)$ — Uniform distribution

$\mathcal{N}(\mu, \sigma^2)$ — Normal distribution with mean μ and variance σ^2

i.i.d. — independent and identically distributed

Linear Algebra

c — A scalar

$\mathbf{x}$ — A vector

$\mathbf{A}$ — A matrix

a_{ij} — Matrix element in row i, column j

$\mathbf{A}^{\mathbf{T}}$ — Transpose of matrix $\mathbf{A}$

$\mathbf{I_n}$ — The $n \times n$ identity matrix

$\mathbf{A}^{-1}$ — Inverse of matrix $\mathbf{A}$

Statistics

μ — Population mean

$\bar{x}$ — Sample mean

σ — Population standard deviation

s — Sample standard deviation

s^2 — Sample variance

s_{ij} — Sample covariance

r_{ij} — Sample correlation

H_0 — Null hypothesis

H_A — Alternative hypothesis

$\hat{\theta}$ — Parameter estimate

$\mathcal{L}(\boldsymbol{\beta})$ — Likelihood function

$l(\boldsymbol{\beta})$ — Log-likelihood function

Datasets

$\{\mathbf{x}_i\}_{i=1}^n$ — A dataset

$\{(\mathbf{x}_i, y_i)\}_{i=1}^n$ — A labeled dataset

$(\mathbf{x}, y)$ — A labeled data point

$\mathbf{x}_i$ — The ith data point

1

Introduction

1.1 What Is Data Science?

There is a great need for data scientists, but at the same time, there is also a great need for people to understand what the field is about. For many, data science is a completely unfamiliar term. Many educators and universities wonder how it is different from more mature fields like computer science and statistics. This might partly explain why some schools have data science in the computer science department, while others might have it within statistics, engineering, a school of information, or as a standalone school (as at UVA).

I won't ask for trouble by stating where data science should live at a university, or how it should be implemented. I do want to offer a definition of the field, to provide some basis for understanding:

Data science is a multidisciplinary field that explores trends and relationships in data, and solves problems using information in data. It draws heavily from statistics, mathematics, and computer science.

Beyond these fields, as data is gathered from people and other living beings and used to make decisions, important questions arise in ethics, law, privacy, and security. Business questions arise, such as what data is valuable, which algorithms to build, which datasets to use, and how data products can fairly serve others. The practice of data science touches so many areas, which makes it broad, complex, and exciting.

Data science is intertwined with computer science: computations are done by computer, data is often stored in databases and similar structures, and *algorithms* (recipes) are used to automate tasks.

1.2 Relationships Are of Primary Importance

Data science explores, identifies, and harnesses persistent relationships between variables. These relationships are used for explanation and prediction (referred to subsequently as prediction going forward; this will be clarified when the distinction is necessary). The quantity being predicted will be called the *target variable*. The variables used for prediction will be called the *predictors*. As data science has many groups of participants using specific terminology, elsewhere the target might also be called the *response variable* or *dependent variable*. A predictor might also be called an *explanatory variable, independent variable, factor*, or *feature*.

1.3   Modeling and Uncertainty

Many of the machine learning models in use today, such as regression models, trace their origins to statistics. In 1885, Sir Francis Galton introduced the idea of regression when he examined the relationship of fathers' and sons' heights [2]. Statistics rests upon probability theory, which is a branch of mathematics.

This leads to an important point that I want to make early:

Much of data science rests on systems and methods which are inherently uncertain.

When predictive modeling is used, the model will make assumptions. It will take variables as inputs. The inputs may be subject to uncertainty in measurement and *sampling error*. For example, drawing again from the population might produce data with different properties and relationships. Finally, where we cannot explain a relationship with equations, we might turn to simulation. The simulation will face uncertainty. This is not to say that we shouldn't simulate, model, or do data science. This is to say we must do it carefully, we must understand the assumptions, and we need to properly test and monitor the products.

1.4   Pipelines

1.4.1   The Data Pipeline

For data to be useful in a system, it often needs to be ingested, processed, and stored. Processing steps generally include:

- Data extraction, which parses useful parts of data from volumes of ingested data

- Data cleaning, such as detecting and fixing problematic values

- Combining datasets to provide a more comprehensive view of the subject matter

- Data transformation, such as converting units of measure and normalizing values

The structure that handles this end-to-end process is called a *data pipeline*. To advance the idea beyond just a blueprint for handling the data requires developing code ... and usually quite a lot of it. Since it can be such a heavy lift to build and run a reliable, efficient pipeline, specialized software applications have been developed for this purpose. For example, Amazon Kinesis can ingest and process streaming data in real time.

1.4.2 The Data Science Pipeline

A second pipeline that arises in data science work is the *data science pipeline*. One possible sequence of steps is shown in Figure 1.1, which is loosely based on the Machine Learning Lens of the AWS Well-Architected Framework [3]. Movement generally flows from left to right, but there are many cycles as the data scientist iterates to build better models. The steps consist of clearly defining and understanding the objectives, framing the objectives as an analytical problem, collecting appropriate data, preparing the data for modeling, iteratively building and evaluating models, deploying the winning model, monitoring performance, communicating the results, and adjusting the model as needed to meet objectives. If the model does not meet objectives, the steps will be repeated. As modeling proceeds, steps in the process might be repeated several times, such as collecting more data, preparing and exploring it, building additional models, and evaluating them.

The data science pipeline may use the output of data pipelines mentioned earlier, but it is a separate process where the purpose is to produce a useful model. A data scientist will typically spend a large amount of time processing the data, testing different predictors, evaluating the results, and analyzing the errors.

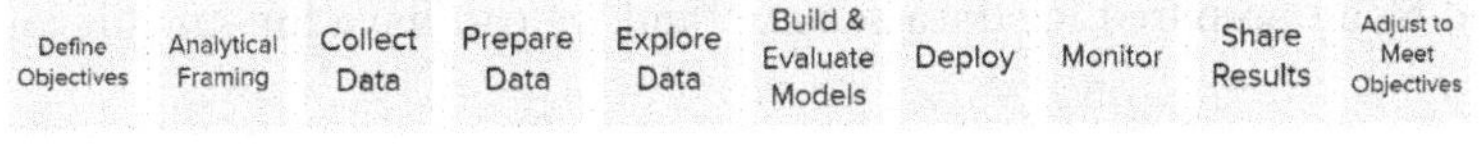

Define Objectives	Analytical Framing	Collect Data	Prepare Data	Explore Data	Build & Evaluate Models	Deploy	Monitor	Share Results	Adjust to Meet Objectives

FIGURE 1.1: The Data Science Pipeline Process

1.5 Representation

An essential step in preparing data for modeling and analysis is to represent the data in a useful form. A data representation needs to include enough detail to solve the problem at hand. It also requires a data structure that is compatible with the algorithm.

Taking natural language as an example, we might wish to classify the sentiment of a review such as:

"Their bar makes a mean dirty martini"

To do this calculation, we need to convert the words to numbers, but what are the right numbers? A good representation will strike a balance where the important information is retained (such as the relationship between the words) and the unnecessary information is discarded (such as the word "Their"). This is not an easy problem to solve, but good solutions do exist, and the right representation is needed for solving the problem.

1.6 For Everyone

Data science is for everyone, and done properly, it can benefit everyone. Predictive models are in use throughout the world at all kinds of organizations, solving a myriad of problems. We can teach the concepts and technologies to the workforce of today, and our students – the workforce of tomorrow – so that they can participate in data science.

I have worked with individuals who selected their career fields to entirely avoid math and statistics. Later, they found themselves designing surveys, compiling the results, and reporting them to upper management. To their surprise and dismay, they were doing statistics and they were unprepared. We all need to be prepared for data science, and we can have fun on this journey.

1.7 Target Audience

Data and data-driven decision making pervades every aspect of our daily lives. This book is intended to provide a valuable set of tools and an understanding of data literacy, analytics, and predictive modeling. The prerequisites are lightweight: an understanding of algebra including functions. Programming knowledge is not required, but prior exposure to a programming language will be helpful. The target audiences for this book are:

- First- and second-year undergraduate students and advanced high school students

- Professionals working in data or software-related fields looking to learn about data science

- Anyone with the interest to understand more about data and how it is being used in our AI-driven world

1.8   How this Book Teaches Coding

Doing data science well requires writing good code. This book covers the fundamentals of coding, but more importantly, it will help with becoming a self-directed programmer. To make the most of this book, it is best to actively work through the code. This means running the code snippets, reviewing the detailed explanations, trying variations on the code, and investigating errors.

This text uses Python because it is one of the most popular languages for data science, it supports the tasks we need to accomplish, and it is relatively easy to learn. It will teach Python by diving into problems with data and solving challenges. It will not methodically catalog what each operator does, what each control structure looks like, and so forth. There are excellent references for learning in this way, with suggestions given later. Finally, the book includes many exercises across different topics, with the hope that they are interesting and motivating.

1.9   Course and Code Package

Accompanying this book is the content of the introductory data science course targeted to advanced high school students, first- and second-year undergraduates, and others interested in getting started in the field. It consists of agendas that teachers can use for planning and preparation, lecture notes, and coding assignments. Most of the lecture notes and coding assignments are in Jupyter notebooks, which contain rich text, images, links, and runnable code. The notebooks are very useful for teaching and demonstrations.

The course materials were designed to be sufficient for a one-semester or two-semester introductory data science course. Teachers can assign readings and exercises from this book to support the course. A one-semester course might cover chapters 1–6, 8–13, and 16–17. This would give an excellent overview of data science while skipping more advanced topics including logistic regression and clustering.

1.10   Why Isn't Data Science Typically Done with Excel?

Excel and similar spreadsheet applications are packed with features and they can present all of the numbers to the viewer. From my experience in industry, some managers love the visibility that Excel brings. However, it can be very easy to make large mistakes, such as deleting data, failing to update ranges, and sorting only part of a dataset. Extending the analysis, tracking changes, making work reproducible, and automating the process can be a challenge. Learning a language such as R or Python to do data science requires a greater upfront investment, but it is well worth the effort. How can one satisfy the manager who loves to see all of the numbers? After the analysis is finished, the results can populate dashboards, tables, or another format that stakeholders prefer.

1.11   Goals and Scope

This book is a first course in data science. It will teach essential skills including:

- How to use Git and GitHub to work with a code repository

- Coding in Python to process, summarize, transform, and visualize data

- Classical machine learning tasks: regression, classification, and clustering using `sklearn`

- Troubleshooting errors, which are a daily part of the life of a data scientist (sorry!)

Beyond these technical skills, the reader will learn to think holistically about the field, paying attention to critical topics such as:

- Effective communication of technical topics to broad audiences

- Understanding how bias plays a role in data science

- Learning how to incorporate ethical practice into data science

- Data literacy

As the field of machine learning is broad and rapidly evolving, and this book is an introduction, it will not cover all of the popular models. It will stop short of treating neural networks, for example, although they are briefly discussed in the final chapter. References will be provided for going deeper. For a detailed review of machine learning models, for example, see [4].

1.12   Exercises

Exercises with solutions are marked with Ⓢ. Solutions can be found in the book_materials folder of the course repo at this location:

https://github.com/PredictioNN/intro_data_science_course/

1. Ⓢ Explain data science in terms your grandmother would understand.

2. Ⓢ Your grandmother asks how data science is different from statistics. Explain the difference in simple terms.

3. What is something that excites you most about learning data science?

4. Give an example of something that you believe to be random and cannot be predicted.

5. Ⓢ Give an example of something that you believe to be random but could be at least partially predicted. What might you use to predict the outcome?

6. Ⓢ Explain how the data science pipeline is different from the data pipeline.

7. One of the steps in the data science pipeline is preparing data. Give an example of data preparation.

8. Name an industry that you believe does not use data science. Search the web to confirm or reject your assertion.

9. Ⓢ Explain why data literacy is important.

10. Can you think of an example when data was misrepresented with the intent to mislead the reader?

2

Communicating Effectively and Earning Trust

Once you move beyond doing data science for yourself as a hobby, you'll need to communicate your work with others. Data scientists have the potential to be transformative and highly visible, and it is not uncommon for them to present work to leadership, board members, venture capitalists, and customers. Effective communication is so essential that if you don't master it, no paying customer or employer will enlist your technical services for long. That is why this chapter appears before the technical topics. A side effect of learning strong communication skills is better relationship building, which is associated with greater happiness and fulfillment [5].

One of my interviews at a healthcare IT firm was with a seasoned clinician who I'll call Janice. She was apprehensive about my ability to come in and lead a team of data scientists to transform their electronic health record system. To my surprise, she had worked with many scientists before who had developed neural networks to predict health outcomes. Digging deeper, it turned out that Janice had been burned by data science before, and the pain radiated through the organization. This is very common, unfortunately, as data science outcomes can be highly uncertain, and the field has been overhyped. To win over Janice, you'll need to earn and maintain trust. You will need to show that you are competent and responsible, and that you respect and value people.

2.1 Master Yourself

Before you can win over the hearts and minds of others, you need to convince yourself that you're doing the right things for the right reasons. If you're taking an introductory data science course and you plan to go no further in the field, then you have a compelling reason that doesn't require a lot of digging deep: you're doing this to meet a requirement. My hope, however, is that you fall in love with data science and make it a longer-term endeavor. In this case, having a "why" and revisiting and reinforcing it will be invaluable [6]. When we have a burning reason for doing something, it propels us forward through adversity, and it gives others an entry point to admire our work. This is how great leaders can influence others. As a bonus, when an interviewer says "Tell me about yourself," your "why" will do the important work of telling your unique, important story.

In addition to having a "why," you will need a strong belief in yourself. You may worry about the computer science requirements of the field, public speaking, or some other aspect of data science. From the start, keep an open mind, stay positive, keep a list of topics for exploration, and gradually chip away at the challenges. Even for accomplished professionals, feelings of self-doubt and not deserving a "seat at the table" are very common. The phenomenon is so prevalent that it has a name: *imposter syndrome*. A meta-analysis in the *Journal of General Internal Medicine* found imposter syndrome prevalence rates ranging from 9% to 82% across 62 studies and 14,161 participants [7].

Having some quick wins in your strength areas can help build confidence. For the areas that need improvement, form a plan for moving forward. You may be able to get coaching from a manager or more senior colleague. It is perfectly fine to be honest about things you don't know and ask for help. Finally, a mentor can help greatly. People in the highest ranks of large organizations use mentors to share ideas, identify weaknesses, and identify opportunities for growth.

Let's imagine now that you've been working in data science for a few years and you feel confident doing tasks like coding, preparing data, and building machine learning models. While these functions may not change, the underlying tools and techniques will constantly change. I mention this because I want you to be mentally prepared for this certainty. Data science is a field that constantly evolves, as it is driven forward by technology and changing customer demands. If you can accept the challenge of learning new things, and the idea that you're never done, this will keep you relevant and engaged. For support and resources, I encourage you to take part in activities that you enjoy, such as joining a data science community, attending conferences or talks, or reading online blogs.

Finally, think deeply about what you want to accomplish in your lifetime, put in the work, and have courage. Here is one of my favorite quotes from Jonas Salk, the developer of one of the first successful polio vaccines:

"There is hope in dreams, imagination, and in the courage of those who wish to make those dreams a reality."

2.2   Technical Competence

In my first job at Merrill Lynch, my manager had an uncanny knack for looking at a table of numbers and finding errors. He didn't need to see the code or pick through the details. He used *heuristics*, or rules of thumb, to examine results critically. For example, he might realize that a column of numbers should sum to one. Great managers and leaders can do this, so it's best to be careful and thoughtful. As you do the simple things properly, you will be trusted to do harder things.

A great data scientist needs to be a thoughtful "trail guide." Like any technical field, data science uses abstraction, assumptions, complex ideas, and jargon. Data and models have limitations and weaknesses that may render your careful analysis useless. To earn trust, make everything crystal clear to your audience. State assumptions and limitations up front. Explain why the approach is useful in spite of the assumptions and limitations. Add a key for acronyms, define the variables, and briefly explain the metrics. Are you summarizing customer satisfaction with a score on a scale of 1–10? Explain how this is measured. If possible, test the assumptions and measure their impact, and try things in different ways.

When we review the work of others, we tend to look for logical fallacies. Methodology issues early in a presentation can quickly eradicate the interest of the audience. "Wait ... this dataset is missing a key group of individuals. In that case, nothing else matters in this analysis." When sharing our work, it is ideal to avoid making such errors, and to anticipate tough questions by addressing them in the work. An effective strategy that Amazon uses when developing a new product idea is to write a PR/FAQ, which is a combined press release and question/answer section. The question/answer section thinks around corners and gets ahead of reader questions and objections.

2.3   Know Your Audience

As an advisor and consultant, I work with customers to conceptualize, plan, and implement applications of AI and machine learning. This includes working with their leadership to prioritize initiatives, plan a detailed proof of concept (POC), help with implementation, and report out the findings. From project start to finish, the audiences will change and I need to be thoughtful on how best to lead discussions. As a technical person, I need to avoid my temptation to talk about deep neural networks with a CEO, for example. I need to "know my audience," so let's get to know them a bit.

My meetings with the leadership team may include the CEO, the Chief Product Officer, and the Director of Engineering. In these discussions, the focus will be at a strategic level; it is not about fighting fires to solve a small, immediate problem. We talk about things that can have a large impact on the business, such as creating a new product. If we discuss a metric, it will most often be return on investment (ROI), which reflects expected revenue and capital investment. The questions are focused on *what* should be done, and not *how* it should be done. Of course, machine learning is part of the *how*, but we won't get into these details with this group. First, this is likely not the right team to answer the *how*, and second, it is generally more effective to brainstorm and think through the *how* later.

After identifying the right project, I meet with a technical team that has knowledge of the data and process. This might include an engineering manager and a data science manager, as well as lead data scientists and data engineers. At this level, we can talk about the *how* in great detail, such as nuances in the data, the strengths and weaknesses of various models, and performance metrics. Not every technical team has data scientists, and I will need to explain things appropriately and educate the group as needed.

After the POC is completed, I report back to leadership, and the technical leaders will join this discussion. At this point, I have an understanding of the business problem we are trying to solve, the approach taken in the POC, and the results of our efforts. It is possible to talk in great detail about the approach and performance metrics, but this likely won't be appropriate with this group. If I feel the details may come up in discussion, then I will put slides in an appendix.

When meeting with leadership, it is extremely useful to put an *executive summary* up front. These meetings can sometimes spur lively discussion at the beginning, and the presenter may not get far into slides. The summary will make all of the essential points. When presenting results, it is best to keep it brief, and provide clear definitions. Help the audience understand why the results are valuable, in business terms when possible. If a case will be made for replacing their current model with a newer model, it is best to quantify the impact in dollar terms when possible. Bear in mind that leaders may have direct experience using "new, improved" models that didn't deliver what they promised.

The guiding principle for meetings and presentations is to align on the agenda. When possible, come to agreement on topics of interest in advance. It also helps to learn about the audience members before the event: their backgrounds, roles, and responsibilities. This will make for a more engaging, valuable discussion.

2.4   Tell Good Stories

If you have ever found yourself at a presentation where you were overwhelmed by a large table of numbers, you understand the value of proper context. A table of numbers cannot be expected to support itself. It is not enough to make a decision. A more effective way to communicate data science findings is to craft a story with a narrative and visualization, grounded in data. We will look at an example, but first let's start with an effective storytelling approach used by product teams.

Product teams make stories a central part of their process. A common template looks like this:

> A [description of user] wants [functionality] so that she can [benefit].

This clarifies the persona involved and how the work will be helpful. A specific example might be:

Content Discovery: User Story

> A *seventh grade math teacher* wants *a way to easily search for relevant content* so that she can *more quickly plan her lessons.*

Data science can often provide the functionality that a product requires. For the content discovery story, a recommender system might be added to help the teacher. Since the audience may be broad, we will mention the type of model, but will refrain from getting into details. Let's add to the story, providing layers about the data and the model.

Content Discovery: Data

> We can leverage our database of information about the teachers and the content. The content includes things like articles, lessons, and videos. The content is tagged with things like the appropriate grade level, the learning objectives, and the subject area. We will know which grade levels and subjects are taught by the teachers. We also have all of the interaction data: each time a teacher selects a piece of content, we have the teacher identifier, content identifier, and the timestamp. We know *who* selected *what* content *when.*

At this point, I should mention that this would be the ideal data for this kind of system. As you might imagine, many schools may not yet have this kind of information. Let's continue with the model layer.

Content Discovery: Model

> We trained a recommender system on all of the data from the past school year. The model was rigorously tested and we measured latency (the time required to get back recommendations), and the quality of the recommendations. The model is stored securely in production. Teachers can click a button in a web browser to see a list of useful content.

This layer touches on important considerations like data security, speed, and relevance of output. The audience will probably be curious if the results look sensible, and the story should include examples like this:

Content Discovery: Usage and Results

> Jamie teaches seventh grade math. Her students learned about representing numbers in decimal form. She is planning her next lesson, which will require students to learn about representing numbers as percentages. She logs into the learning management system with her credentials and clicks the button *Get Next Lesson.*
>
> The recommender system returns a list of 20 suggested titles with keywords and learning outcomes. Here are the top three titles:
>
> 1. Pterodactyl Percentages
> 2. 100%! Converting Decimals to Percentages
> 3. Blue Horseshoe Loves Percentages

Note that this was a small example, and I picked a recommender example. The story can be crafted to meet the needs of the audience, with additional layers as appropriate. Let's put all of the layers together to see a compelling story:

Content Discovery: Complete Story

> **User Story**
> A *seventh grade math teacher* wants *a way to easily search for relevant content* so that she can *more quickly plan her lessons.*
>
> **Data**
> We can leverage our database of information about the teachers and the content. The content includes things like articles, lessons, and videos. The content is tagged with things like the appropriate grade level, the learning objectives, and the subject area. We will know which grade levels and subjects are taught by the teachers. We also have all of the interaction data: each time a teacher selects a piece of content, we have the teacher identifier, content identifier, and the timestamp. We know *who* selected *what* content *when.*
>
> **Model**
> We trained a recommender system on all of the data from the past school year. The model was rigorously tested and we measured latency (the time required to get back recommendations), and the quality of the recommendations. The model is stored securely in production. Teachers can click a button in a web browser to see a list of useful content.

Usage and Results

Jamie teaches seventh grade math. Her students learned about representing numbers in decimal form. She is planning her next lesson, which will require students to learn about representing numbers as percentages. She logs into the learning management system with her credentials and clicks the button *Get Next Lesson*.

The recommender system returns a list of 20 suggested titles with keywords and learning outcomes. Here are the top three titles:

1. Pterodactyl Percentages
2. 100%! Converting Decimals to Percentages
3. Blue Horseshoe Loves Percentages

Now we have an end-to-end story that can be more readily consumed than a table of numbers or a graphic. To be sure, relevant numbers and graphics should be included to support the narrative, but they cannot replace the narrative. The story clarifies the persona (the *who*), the functionality (the *what*), the benefit (the *why*), and the plan (the *how*). It also explains the purpose of the model, the problem that it solves, and how it goes about solving the problem. The story is self-contained, and for those who want to know more, additional information can be provided.

2.5 State Your Needs

Regardless of how good you are and how well you know yourself, there will be things you will need to be successful in a data scientist role. As long as work continues to flow to you, the needs will come with them. Even during slow times when people aren't asking something of you, it will be helpful to take initiative, explore, learn, and look for work.

If you are new to a company or team, you will have several basic needs like hardware, software, and network access. You will need to know your team's pattern of communication. Perhaps for quick updates, Slack is preferable, while emails are encouraged for detailed messages. Some companies have a meeting culture, while others avoid them whenever possible. You will need to know where colleagues store things like code, data, and documents. As you are given projects, you will need ongoing support from key people.

Getting stuck at work can be scary, particularly for a new employee. Data scientists can get stuck a lot, because coding issues and data problems crop up regularly. Whenever possible, I encourage you to figure things out for yourself, as this can be gratifying and it shows independence. However, if you feel that you're struggling or you have hit a wall, then do not be ashamed to ask for help. This will help people to trust you, actually, because they will get the message that if you are stuck, you will work to get unstuck. This is preferable to people counting on you to get work done, waiting for weeks as you struggle in silence, and ultimately having nothing delivered.

In some of my early roles, I faced steep learning curves and my managers were very busy people. They usually couldn't provide help at a moment's notice, so I asked them to suggest how I could get help. Answers varied, of course, but one thing that helped was collecting all of the *blockers*, or places where I was stuck, and thinking through them. I found that for some, I could break them down into subproblems and then tackle the overall problem. For other problems, there was information that I needed to move forward. This habit shortened my list of blockers and it made me more independent. As people freed up to help (typically after the financial markets closed), we had fewer things to work through. The second big thing that helped was developing an internal network of people in similar roles. There was a good chance that someone near my level could find a little time to offer help.

The current environment is great for crowdsourcing your questions. Instant message boards like Slack or Teams are conducive to posing questions to groups of people and getting answers in a timely fashion. If you have a coding question, there is a great chance that web resources like Stack Overflow will have an answer. In the other direction, I encourage you to share your knowledge and help others when you can.

Beyond getting help for basic setup and staying productive on projects, you should also think about what you need to grow and to love your work. This might include challenging projects, headcount, a promotion, or a raise. Whatever the request may be, just be sure that you really want it, that you are ready for it, and that you can document why you deserve it. Some of these things come with more responsibility, or different responsibilities. For example, changing from individual contributor (IC) to manager will require a different set of skills. ICs need to produce work, while managers need to motivate and coach others to produce the work. Not every IC will enjoy being a manager! Some data scientists remain ICs for the duration of their careers.

2.6   Assume Positive Intent

It is very hard to know what people are thinking at any given time. Making this more challenging is remote work and communication channels such as email and instant messaging. We can lose vital information including body language, inflection, and tone. At the same time, many data science roles involve working with people under challenging conditions that include time pressure, uncertainty, and changing requirements. All of this poses difficulties as we try to maintain positive relationships with coworkers.

As director of an online master's program in data science, I watched many of my students struggle to balance a demanding degree, family responsibilities, and a full-time job. Several projects involved group work, and it was difficult for teammates to work at a constant pace. This was the first time that I heard mention of the term *positive intent*, which suggested that students assume that others are trying their best and mean no harm, particularly in ambiguous circumstances. As one might imagine, it can be very easy to interpret a Slack message negatively (especially when it's missing the right emoji!). Moreover, once things take a negative turn, escalation can quickly follow. The advice was very powerful, because assuming positive intent brushes off small things that may mean nothing. It means giving the benefit of the doubt to the other person, and this can induce empathy.

In *Crucial Conversations* [5], the authors suggest that we invent a story to explain behavior. When things don't align with our wishes, the story is often negative. We may assume a manager has a low opinion of our capabilities, or that a colleague doesn't like us. The tragedy is that the damaging story may not be correct. As an example, I had an excellent summer intern who was very driven and results-oriented. I'll call her Christine. As the summer was ending, a colleague mentioned that Christine loved working for my team and that she was hoping she could return as a full-time employee when she graduated. However, since she wasn't offered the option to return – some of her friends working for larger firms had received these offers – she was feeling increasingly discouraged. She made up a story to explain what was happening, and this story included the message that her work was not valued. Of course, this was far from the truth. The truth was that we worked at a startup which was long on work and short on process – our timing simply didn't align. I was very grateful to my colleague in this instance, as it provided me the opportunity to eradicate the negative story. We made Christine an offer, and she still happily works for the company.

My story was about an intern not asking for a full-time position. This specific scenario may not apply to you, but a more likely scenario that may apply during your career is that a pay raise or promotion is in order. A sad but true fact is that many companies are not actively thinking about how to give more money to their employees, even when they highly value their efforts. This is another example where assuming positive intent and having a conversation, even when uncomfortable, is preferable to inventing a negative story. In the space between conversations, we build up stories and feelings to explain differences between reality and expectations. Asking for things, including honest conversations, will help with alignment.

2.7 Help Others

I want to briefly make the case for helping others through knowledge sharing and displays of empathy. Earlier, I mentioned that you will need help from others, so it's only fair that I ask that you help others whenever possible. Helping others is not only good karma, but it will increase your satisfaction and value to the organization. Ultimately, we feel good when we can be helpful, and it promotes a message of respect and care for our teammates.

There are many ways to be helpful. If you see someone struggling and you can help them, please go ahead and offer help before they ask. Some organizations form groups of employees with specialized expertise helping on a common goal. At UVA, there was a desire to scale the number of online programs. A team of leaders from across the university would regularly convene to share best practices. Such groups provide several benefits, including opportunities for networking and accelerated learning.

I have worked in all kinds of organizations, from startups to massive enterprises, and from competitive places to cooperative ones. Of course, everyone is entitled to their opinion, but I find that cooperative organizations promote stability, support, and well-being. You might ask why people choose not to help others. There can be several reasons, but two common factors that I have found are the incentives from leadership and a *scarcity mindset*.

People generally respond to incentives. Many salespeople carry quota, such as a certain amount of revenue they need to bring in each quarter. The quota is a dangling carrot that leads to planned behaviors. If a team of data scientists are told they need to mentor more junior data scientists before they can be promoted to managers, then you can be sure that the ones that want this role will respond to the incentive. When leadership rewards workers for certain values, and the workers agree with these values, they tend to propagate them.

Now let's think about a scarcity mindset and an *abundance mindset* [8]. If you feel that you have a great deal of some resource (e.g., money, opportunity, knowledge), then you might feel generous with it. This is the abundance mindset. You might donate money or train a junior employee. If you worry that you don't have enough for yourself or your family, then you will be reluctant to give it away. When I worked on Wall Street, I met several people who were afraid of colleagues aiming to take their job. As a result, they were selectively helpful, as they feared that indiscriminately helping people would make them replaceable. In certain situations, this seems understandable. In general, however, playing defense is a strategy based in anxiety, and anxiety is draining. I encourage playing offense: learn as much as you can and focus on your growth. Learning will yield an abundance of knowledge, and sharing this knowledge will make teammates better and deliver value to the organization.

2.8 Take Ownership

This section is not about conquering territories, bending people against their will, and running a tyrannical empire. It is about earning the privilege to manage and lead things by taking full responsibility for outcomes. Management entails keeping things running smoothly, generally by following established practices, while leading entails breaking new ground and providing a vision for the future. Both management and leadership are essential, and they both require ownership.

A guiding career principle to keep in mind is that your boss likely wants a promotion, and to get a promotion means delivering great outcomes. When you are interviewing with a hiring manager, and when you are on the job, the manager is asking herself: "Will this person be an asset to the team?" If the answer is "no," then the probability of getting the job or the promotion will be low. When the boss assigns a project to a person or a team, she needs to rest assured that it will be completed within the timeframe with proficiency and a good attitude. The basics of project management will apply: the project members will need to know the tasks to complete, the roles, the deliverables, and the deadlines. You don't need to raise your hand on everything, but you need to deliver on your promises.

To get in the mindset of an owner, you might need to adopt some new ways of thinking. First, you can start by thinking that if you don't do something, it won't get done. This may be accurate for some tasks, particularly if you are the only person with the required skills. Second, you can assume responsibility when things go wrong. If a teammate makes a mistake and customers complain, you can put the blame on yourself. The manager didn't give clear instructions? You can say this was your fault for not asking for clearer instructions. Is this reasonable? Perhaps, and taking the blame will make you more likable than directing blame at others. Nobody likes to be blamed for things. If you truly want to show ownership and win people over, this can be an effective approach. An excellent book on the topic of ownership is [9].

As you increasingly think like an owner, master yourself, build technical competence, and improve the other skills we've studied, you will feel ready to own things. Raise your hand and ask for ownership. You might get the opportunity to manage small projects or mentor a junior teammate. As you see success, you might be given larger roles, such as building and managing a team or a product. This will bring responsibility and a likely sense of satisfaction.

Once you are an owner, you need to do things to stay an owner. You might need help from others. To get the best from others, people will need to like you, understand your vision, and trust that the mission will be successful. You will need to do your part to clearly communicate what is needed and why, develop the talent of your direct reports, and exhibit fairness, among many other things. There are many excellent books on leadership strategies and ownership, and I highly recommend [10] for going deeper.

2.9 Chapter Summary

Data scientists need to be strong communicators, as their efforts can drive high-value projects of interest to leadership. One challenge is that your teammates and audiences will likely be a mixture of technical and non-technical personas. Earning trust goes beyond technical mastery and must include effective storytelling, breaking down material for a diverse audience, and clearly stating assumptions and limitations.

A prerequisite for influencing others is to believe in the value that you bring. You will need to feel that you belong and that you are compelled by the work. From there, building your skills and brand are important. When presenting, it helps to know the audience. If possible, research the attendees in advance. This will allow you to deliver your message appropriately, with the right content and level of technical material.

When reporting on results, it won't be sufficient to show a table of numbers or a graph. These are important, but a story that integrates the numbers and graphs will work better. A data scientist should craft a narrative which begins with the problem being solved and the value it will deliver. Then it can move into a discussion of the technical material at an appropriate level: the data, model, and metrics. Finally, stakeholders will want to see clear, correct results.

Throughout your career, there will be things that you need. It may be something simple, like an extra monitor, or something bigger, like ongoing training. When the need will make you more effective in your role, you should feel empowered to ask for it. If it is something like a promotion, you should be able to provide justification.

Working relationships need regular communication to stay healthy and productive. In the absence of conversation, or when social cues like voice and body language are missing (e.g., when sending messages on Slack), it can be easy to feel slighted. To avoid misunderstandings and reduce stress, it can be helpful to assume that people mean well. A common pattern is for people to invent damaging, personal stories that explain why a negative outcome occurred. Rather than make up a story, try to have a conversation and work to understand the reality of the situation.

Environments where people exchange ideas and provide support are enriching. I encourage collaboration and helping others when possible. Rather than playing defense and hoarding knowledge, commit to lifelong learning and sharing knowledge. This will make teams better and deliver value to the organization.

It can be very empowering to be responsible for things. To gain the privilege of responsibility, you'll need to prove that you can own things. Show people that you can be trusted to deliver on promises. Assume that if you don't complete the task, no one else will. As your responsibilities grow, you will need help from others. Provide them with proper context and motivation to help you. This means showing others that you genuinely care about them, and that you want to help them grow.

I encourage you to build and refine your communication skills, which will pay dividends in both your career and personal life. In the next chapter, we will turn to the big picture of planning a data science project. The planning exercise helps to increase the odds that the project will be successful. This is essential, as data science projects are often complex and require cross-functional teams.

2.10 Exercises

Exercises with solutions are marked with Ⓢ. Solutions can be found in the `book_materials` folder of the course repo at this location:

https://github.com/PredictioNN/intro_data_science_course/

1. What gets you out of bed in the morning?

2. Discuss a time when you made a technical mistake. What happened? Were you able to earn back trust?

3. Ⓢ Give an example of something that should be included in a presentation to an engineering leader but not to the CEO.

4. Ⓢ Why is it important to include an executive summary at the start of a presentation to leadership?

5. Ⓢ What should be included in a data science story?

6. Ⓢ True or False: A data science story should begin with a table of numbers.

7. Ⓢ Indicate if the following product story is a good template, and explain your reasoning:

   James is a radiologist. He needs a fast, accurate system that can review chest x-rays for tumors.

8. Ⓢ You've just started a new data science role. What is the best way to get unstuck at work? Select the best answer.

a) Wait until someone comes to offer help.
b) Keep trying until you eventually find the solution. It's best not to waste the time of your colleagues.
c) See if you can quickly solve the problem yourself, and then ask for help if needed.
d) Ask your manager or nearest colleague each time you have a question. There is no shame in asking for help when it's needed.

9. Ⓢ Which of these selections are examples of blockers? Select all that apply.

a) The necessary dataset is not yet available for the project
b) A different engineering team needs to build a feature before the next team can begin their part of the project
c) The development team meets each morning to discuss the project status and plan for the day
d) A manager needs to sign off on the project and he is not responding to requests

10. Explain why it can be helpful to assume positive intent.

11. Explain the scarcity mindset.

12. Ⓢ Provide an example of how you can demonstrate ownership of something.

13. Why do you think this chapter appears in a book on data science?

3

Data Science Project Planning

This chapter discusses the earliest steps in the data science pipeline, which consist of planning activities. The first step is gaining clarity on the problem that needs to be solved, the requirements, and the constraints. The problem is often stated as a business objective, such as increasing the profitability of a product.

Next, the data science team needs to state the problem in an analytical framework. For example, a specific machine learning technique might work well, combined with a loss function and metrics. The executives won't need this level of detail, but the team doing the work will use this blueprint to move forward.

Finally, a plan for data collection needs to be in place. The data might already be available, or it might need to be collected or purchased. If there is no way to get the required data, then the project won't be able to move forward, so it is important this is discovered early.

3.1 Defining the Project Objectives

In traditional software development projects, software teams are accustomed to spending a lot of time and attention writing requirements. The development work won't start until this stage is completed. Data science should follow a similar disciplined process. It may be tempting to launch into analysis, but without a process in place, the project can diverge from the problem that needs to be solved. Additionally, the initial planning can help clarify objectives and identify resources that are needed, such as data, tools, and key people.

Data scientists might work on a mix of internal projects for the company and external projects for a customer. In either case, it is essential for the objectives to be defined and clear to all stakeholders. For example, at one startup tech company, my team was tasked with building a system that could detect objects in images. When we asked the customer about the level of performance it would need, they responded with a required predictive accuracy of 75%. We got to work, building computer vision models and a massive dataset of images. As the project progressed, the accuracy target kept increasing ... 80%, 85%. Even worse, when we clarified the metric, it turned out that accuracy wasn't what they cared about! The customer was adamant that when the model detected the object, it needed to be right. They were not concerned about cases where the model did not detect the object. This meant that the metric was actually precision and not accuracy (we will learn more about metrics later). Ultimately, my team was not speaking the same language as the customer, and thus we were not aligned. After six months of work and additional changes to requirements, the project was halted.

I shared this story because it was a painful experience and a common one. It also could have been prevented. From that experience and other lessons learned, I developed a planning questionnaire for data science projects. I asked that the questionnaire be filled out as a team before the project would start. This required some skin in the game from the person requesting the work, which seemed fair as the projects would often consume a lot of resources from my team. It was well worth the effort, it turned out. The upfront planning and development of a common language helped improve the odds of data science projects succeeding. Next, let's look at some of the things in the questionnaire.

3.2   A Questionnaire for Defining the Objectives

This set of questions is typically discussed at a meeting with key stakeholders. The attendees are generally from product, engineering, data (when this is a standalone team), data science, and the line of business.

Q1: What is the problem that needs to be solved, and why is it valuable?

This should make the problem very clear, and it should help the requestor understand if it's really something that should be done. Example objectives might be to increase revenue for the flagship product, or to lower the cost of customer acquisitions.

Q2: What data is needed and what data is available?

As mentioned earlier, if the data isn't available to support the initiative, it will never get off the ground. This question is essential for taking inventory of the data, and strategizing on how necessary data might be collected.

Q3: Who are the right stakeholders and what will be their roles?

For a data science project to succeed, it needs support across many teams. There needs to be clarity on the people, their roles, and how much they can support the project. For example, if Jeff is the sole subject matter expert and he's tied up for the next six months, then this project likely can't start right away.

Q4: How good does predictive performance need to be, and what are the important metrics?

If the work is for a customer, they might already have a model and a performance benchmark in mind. They might say that their current model has an area under the ROC curve of 80%, and the new model needs to at least exceed this threshold. Conversely, the customer might not have metrics in mind, and you might need to educate them. It can be difficult to get firm numbers, but I encourage you to push for them and discuss different metrics to get clarity.

Q5: Is there a system in place for running and monitoring models in production?

A member of the engineering team should be able to help answer this question. Assuming the model meets the performance objectives, the next step will be to deploy it into production. For this to happen, there needs to be a reliable system available to support data ingestion and processing, model scoring, and passing results to users. Model scoring, or *inference*, is the process of running new data through the model to get predictions. Passing results to users is typically done with an *application programming interface*, or API.

Details to consider include how quickly results need to be returned (*response time*), how available the system needs to be (*uptime*), and where the data and results will be stored. A system required for real-time scoring will need high uptime and low latency, while a system returning daily batches of results won't have these requirements.

> Tip: Response time is the elapsed time from when the request is made to when the client receives the first byte of the response.
>
> Uptime is the percentage of time that the system is available and working properly. It is usually expressed in 9's, as in "six nines" for 99.9999%. This translates to less than 32 seconds of downtime per year.

If a system isn't readily available, then the discussion can include how such a system might be provided. There are several cloud providers offering low-cost, readily available services including AWS, Google Cloud, and Databricks.

Q6: Is there sensitive information in the data?

If there is sensitive information in the data, then additional steps and extra care will likely be required. For example, the data might need to be stored and prepared on a specific machine for conducting the work. Parts of the data might need to be masked before colleagues can view it. The owner of the data should be able to help answer this question.

Q7: Does the model need to be interpretable?

For some use cases, the model can be treated as a black box. Perhaps as long as the prediction errors are small, nobody is concerned with how the model works or why it arrived at the output. This is increasingly rare, however, because at some point, all models degrade and then stakeholders will want to see justification for the output.

More commonly, predictions will need to be explained to people. For a model that recommends a treatment protocol to patients, a doctor will want to know why 20 mg of medication A is the best course of action. An insurance regulator will want to see the predictors in the workers' compensation claim prediction model.

It is important to know in advance if the model needs to be interpretable, because some model types are more interpretable than others. When we learn about modeling, we will cover regression models, which are highly interpretable.

Q8: What are the requirements for compliance?

Some industries are highly regulated, such as finance and insurance. To protect consumers, there are laws specifying how models should be developed and tested, and how data should be collected and treated. The stakeholders should have awareness of relevant laws to remain compliant.

Models that fail to remain compliant can lead to financial loss, reputational damage, and job loss. In the financial industry, for example, lending models are regularly reviewed to ensure they don't systematically discriminate against protected groups of people. We will study this topic later in the book.

3.3   Analytical Framing

After the objective of the project is understood in business terms, it needs to be translated to an analytical framework. Consider this example: leadership and product convene and set a goal for next year to increase revenue for their streaming video service. To support the project, historical purchase behavior is available. A working group convenes and the questionnaire is completed.

As a next step, it is valuable to walk through the process step by step. This can include mapping out the customer experience. It will be important to know how customers are interacting with the system, and what data is collected from them. Important events should also be understood, like a large product discount that might have temporarily influenced behavior. Given this information, a high-level flow diagram can be created for discussion. This process will expose complexities and gaps in understanding.

Next, members from the data science team can start to frame out an approach. Suppose the company has a transaction dataset consisting of each product purchased by each customer over the past five years. Given this information, it would be possible to find similar users based on their purchases. Within the groups of similar users, some users will have purchased certain products, and others might be interested in those products. This suggests relevant product recommendations. The approach we are outlining is a recommendation algorithm called *collaborative filtering*, and it falls in the domain of machine learning. Given this knowledge, the project becomes very actionable because it can be more easily discussed and researched (among people with this specialist knowledge), and there is code available to implement the algorithm. Non-technical audiences will not need to know the details of the algorithm, but it will help to educate stakeholders about how it works, the assumptions, and the limitations.

This short example illustrated that the data science team worked to place the business problem into the framework of a known problem with a readily available solution. This is the preferable outcome, as it saves time when the problem has been solved and there is available code. Of course, this won't always be the case. It may be that the problem to solve has not been solved before. There might not be an existing algorithm, and there might not be available code. In either case, the data science team should sync with stakeholders to discuss which tools are available, and to estimate the level of effort for the project. Leadership might be okay with a large research effort to solve a new problem if the return on investment is expected to be high. Alternatively, there might be an easier, slightly different problem to solve. The data science team should think of different approaches, and try to estimate the effort and feasibility for each approach.

After a feasible analytical framework is determined, the data science team can plan data collection and use. We review this step next.

3.4   Planning Data Collection and Usage

High-quality data will be essential for any data science project. The kind of data needed will depend on the problem being solved. Here, we will discuss problems that involve making predictions. For example, if we need to predict the direction of a stock price, then *time* is an important dimension. In this case, we will need to collect and analyze data over time, which follows a time series framework. The specific data elements should consist of the target variable and any variables that we believe would predict the target. This might include macroeconomic variables and financial variables. The frequency of the data should also be considered, as shorter-term predictions might require higher frequency data.

For the earlier case of recommending products to users, we need three kinds of data:

- User data: At a minimum, user identifiers (*userids*) are required to tell users apart. Metadata such as age and interests may also be helpful.

- Product data: At a minimum, product identifiers (*productids*) are required. Metadata such as category and synopsis may also be helpful.

- User-product interaction data: This captures each user purchase. We need the userids, productids, and *timestamp* (date and time) of each purchase.

3.5   Data Quantity and Coverage

Another important question to answer is how much data is required. This depends on a number of factors, and I will give some qualitative guidelines. First, there are some conditions where there will never be enough data:

- There are no available predictors that can explain the target variable

If there is no relationship between any predictor and the target, then amplifying the amount of data will not solve the problem.

- The labels (target variable values) are not accurate

If the labels cannot be trusted, this will lead to inaccurate results. Collecting more bad data will not improve the model.

- The predictors are not accurate

This presents the same issue as bad labels, as bad data leads to bad models. Next, let's take for granted that the data is accurate and there are variables that can predict the target. Data coverage of the important dimensions is critical. For example, problems might involve time, location, or subpopulations.

Some problems involving time are subject to cyclicality and seasonality. Financial markets alternate between bull markets and bear markets. E-commerce platforms are busier around holiday seasons. A stock price predictor that is only trained on bull markets will suffer underperformance when the market turns. The data used for these problems should cover several full cycles so that models can learn these patterns.

Real estate pricing is partly explained by price per square foot, but this value will vary over time and by location, such as proximity to the beach (glorious!) or highway (noisy!). Prices may also be driven by season and month of the year. The data should include several full cycles across the locations of interest.

For a model to predict values in a range, there must be instances of the target data over that range. For the real estate problem, if we wish to accurately predict home values between $300,000 and $2,000,000, then we need these examples in the dataset. If the maximum home value in the dataset is $600,000, then any prediction above that maximum should be avoided.

For classification problems such as determining if a check deposit is fraudulent or not, many instances of both fraud and legitimate deposits are required. Making this more challenging, fraud is often rare. Sampling methods such as reusing the fraud cases can sometimes improve the predictive capability of the model, but real data is preferable to synthetic data.

For object detection problems, the quickest way to get started is often to find a pre-trained model that has been open sourced.[1] One source for models is the Model Garden for TensorFlow [11]. Whether building an object detection model from scratch or fine tuning a pre-trained model, the dataset should include cases where the object is in the image, cases where it is not in the image, and edge cases that could confuse the classifier. For example, if the model is looking for pools, an edge case might be a beach, as this might cause misclassifications.

To summarize this section, it is important to collect data across the relevant dimensions for the target values of interest. Including edge cases is recommended where possible, as this can help reduce errors. Notice the lack of a simple quantitative answer, such as "collect one thousand rows of data." The question of minimal data sufficiency requires a nuanced answer. Every dataset is different. However, if you are working on a problem similar to one you have already solved, then the solved problem can offer guidance. Also, once some data is available, it can be explored to understand the predictive power of the variables.

3.6 Sourcing Data

Once there is an understanding of the data that is needed, the search begins! In the ideal case, the required data has already been collected and stored, and it is a matter of gaining access to that data. Depending on the systems in use and privacy requirements, the steps will vary. For example, if the data contains sensitive information, it may first need to be deidentified. If the data is divided across many different machines, or servers, then access may need to be granted to each server. After gaining access, the data may need to be collected from each server.

---

[1] Pre-trained models can be useful for many other tasks requiring massive datasets, such as language tasks.

For any pre-existing data that will be used, its *provenance* should be understood. Users will want to know how the data was collected and what it represents. For example, one of my teams had built a model to predict bankruptcy using a large database of variables. One of the variables had a perfect correlation with bankruptcy. This was very suspicious, and we asked the owner of the data to walk us through the variables. It turned out that the amazing predictor was lagged bankruptcy. Since lagged bankruptcy is something that would not be known in advance, it couldn't be used in the model. Understanding the backstory of the data prevented my team from making a huge mistake.

In the event that the required data is not available, options include collecting the data, purchasing it, or finding some open-source data. For collection, it might be possible to work with teams such as engineering and design to build a collection mechanism. This approach might be possible if relevant data can be collected quickly. If this won't work, then the company might turn to external data.

For certain use cases, there may be free, useful data in the public domain. Another option is to pay for data. There are many third-party data providers such as AtData and People Data Labs that collect, clean, enrich, and sell various datasets. The data transfer is done through an API which allows for programmatic delivery and processing. If there is funding available for data and a provider has something relevant, this may work.

For any dataset, it is important to understand if the data is right for the project. The data should be fit for purpose, and it should be checked for quality, consistency, and predictive ability.

3.7   Chapter Summary

Data science projects require careful planning to increase their chances of success. The planning starts by defining and clarifying the business objectives. We reviewed a questionnaire for defining objectives. This helps stakeholders understand the value of a project and the resources required. To operationalize the data science work, the problem is cast into an analytical framework. This approach allows the data science team to think about sensible models and the necessary data. Next, the data needs to be acquired. It may be that the company already has this data, or it needs to be collected or purchased. If high-quality data cannot be obtained, the project cannot move forward.

Following the planning stage, the data will need to be ingested into the system. The next chapter will briefly review the use of APIs and web scraping for this purpose. We will also review data more broadly from different perspectives including data structures and data processing.

3.8 Exercises

Exercises with solutions are marked with Ⓢ. Solutions can be found in the `book_materials` folder of the course repo at this location:

https://github.com/PredictioNN/intro_data_science_course/

1. Ⓢ Which of these is a business objective?

   a) Minimizing the model recall
   b) Maximizing the revenue of a product
   c) Eliminating bugs from a piece of code
   d) Reducing the runtime of a process

2. Ⓢ Which of these are red flags which should serve as a warning that a data science project may not be successful? Select all that apply.

   a) A clear objective and commitment from leadership
   b) Key stakeholders do not attend meetings
   c) The necessary data is not available
   d) There is no sense of urgency for the project

3. Ⓢ A colleague asks you to build a model for an important initiative. He insists that you get started before planning the project. Which of these actions would be best?

   a) You should get started on the project. If you complete it quickly, you will impress him.
   b) You should report this to his boss. Your colleague is out of line.
   c) You should explain the importance of the planning process. Once the planning is completed, you can both be confident that the project will be successful.
   d) You should explain the importance of the planning process. This will increase the probability that the project will be successful.

4. What should you do if the customer does not provide a benchmark or performance metrics?

5. Ⓢ Explain why latency and uptime are important.

6. Ⓢ A service claims to have "five nines" of uptime. How many seconds of downtime is this per year?

7. Imagine that you have spent six months developing a predictive model for a customer. Suddenly the customer adds a new requirement. Give an example of a requirement that could jeopardize the entire project. Provide a brief explanation.

8. What is analytical framing and why is it necessary?

9. Ⓢ What is user-product interaction data? Why is it important for recommendation systems?

10. A customer has asked you to build a model that predicts if someone will change their phone plan. List some variables that might be useful, and explain your reasoning.

11. Ⓢ Which of these predictive modeling tasks is most difficult?

   a) Predicting the sales price of a beach home when the last three years of sales prices and property information is available for the neighborhood
   b) Predicting the sales price of a beach home at the start of a financial recession when historical sales information is available from a bull market
   c) Predicting the number of checking accounts opened at the start of a new promotion. Data from similar promotions is available from the last three years.
   d) Predicting a student's performance on a final exam given performance on the midterm, quizzes, and homework

12. Ⓢ You are tasked with building a model to predict if a customer will churn (switch to a competitor). You are asked how many rows of data will be needed. Which is the best response?

   a) One thousand rows of data. It is best to be definitive.
   b) Five hundred rows of data for both the case of customers churning and not churning. This will cover each outcome.
   c) You should indicate that you will need cases where the customer churned and did not churn. It is hard to say how many rows are needed, but you will certainly need variables that are predictive.
   d) You should refer to the project that you did for predicting if a stock went up or down. The number of rows used in that project should be sufficient.

13. What is data provenance and why is it important?

14. Ⓢ Your company does not have the necessary data for a project, and you decide to purchase data from a vendor. What are essential things to consider when using vendor data?

15. Give an example where integrating different datasets can help with a project.

4

An Overview of Data

Our working definition of data will come from the Collins Dictionary:

Data is information that can be stored and used by a computer program.

The Cambridge Dictionary gives a similar, but slightly longer definition:

information, especially facts or numbers, collected to be examined and considered and used to help decision-making, or information in an electronic form that can be stored and used by a computer

Nearly all data is now digital, with books holding less than 10% of the world's data. Libraries are working feverishly to digitize and extract information from their physical documents. Digitization has led to improved search and retrieval, and in the age of AI, it enables so much more including classification, translation, and personalization. Data may be produced by people or agents such as robots, sensors, devices, and servers. It may look like:

- Text in the form of social media tweets

- Images taken by a digital camera

- A TikTok video

- Electrocardiogram waves

- Numeric measurements taken by a blood pressure monitor

- Sound waves recorded by a digital audio workstation, or DAW

Now that we have this broad view of data, we will examine it from many different perspectives. This will include understanding the attributes of different kinds of data, the different forms that data can take, how data is treated within systems and applications, and how data can represent (or misrepresent) things.

4.1   Data Types

Each piece of data will have an associated data type, which allows for a set of possible values and operations. A numerical measurement like 121.5 may be compared to another numerical measurement with the > and = operators, for example. The measurements might also be averaged. A word such as "Love" from a social media tweet could be uppercased or lowercased. It would not make sense to uppercase a numerical measurement, and it would not make sense to compute the average two words.

Each programming language will define a set of built-in data types. In Python, the *float* data type would be appropriate for decimal values like 121.5. The *string* data type would be sensible for words like "Love," as strings can hold a sequence of letters, numbers, and punctuation. The two other primitive data types in Python are integers and Booleans. The Boolean data type takes values True and False, such as whether a Twitter account has followers or not. More complex data types can be built from the primitive data types. We will actively work with and study the various Python data types throughout the book.

To summarize, here are the Python primitive data types with examples:

Type	Examples
float	3.14, 2.71, −1.111
integer	−2, 0, 500
bool	True, False
string	'success', 'bag of words'

4.2   Statistical Data Types

Data can also be categorized into *statistical data types*. At the first level of the taxonomy, data may be *qualitative* or *quantitative*. Qualitative data can be further divided into *categorical data* and *ordinal data*. Categorical data can take the form of text or even numeric values. A key property is that it does not have a meaningful ordering. Examples include eye color and zip code. Note that while zip codes are integers, calculations on them are not meaningful (we would not compute the mean of two zip codes). Although categorical data cannot be ordered, it is still highly useful, and it can be included in models. Ordinal data does have a natural ordering system, and an example is letter grades assigned to student exams; a grade of A is preferable to a grade of B, for example.

Quantitative data can be divided into *discrete data* and *continuous data*. Discrete data is data that takes a finite or countable[1] number of values, while continuous data takes an uncountable number of values. An example of discrete data is the number of attendees at different nature preserves at a given time. An example of continuous data is the speed of a fastball pitch, assuming this is measured with perfect precision.

Consider a dataset with several variables, such as eye color, height, weight, and zip code. The categorical variables (eye color and zip code) generally require the same processing steps, while the continuous variables (height and weight) generally require the same steps. For example, one-hot encoding is commonly used to represent categorical data in machine learning models. It is useful to programmatically group the variables by statistical data type. This will make the processing more efficient, and it will keep the workflow organized.

To summarize, here are the statistical data types with examples for each type:

Quantitative	
Type	**Examples**
Discrete	#website visitors, #students graduating
Continuous	depth of a lake, lap time of a swimmer

Qualitative	
Type	**Examples**
Categorical	cat breed, type of a star, occupation
Ordinal	highest grade completed, job category (I, II, III, IV)

Statistical Data Types with Examples

4.3 Datasets and States of Data

A *dataset* refers to related information that can be processed as a unit by a computer. Examples include a table of stock prices over time, or clinical notes on a set of hospital patients.

---

[1] For a countable number of values, each value can be assigned to a number 1, 2, etc., even if the counting process continues indefinitely.

At any given time, we can think of a dataset belonging to one of three states [12]:

- At Rest: the data is in storage, such as a file, database, or data lake

- In Use: the data is being updated, processed, deleted, accessed, or read by a system

- In Motion: the data is moving locations, such as within a computer system or over a wireless connection

A system or application may comprise many dozens or more datasets. At any given time, there may be datasets in each of these three states: new data is ingested into the system, mature data is stored in a database, and a user is updating a portion of data. This may be a very dynamic process. We will learn more about each of these states later when discussing data storage and processing.

An important consideration in each of these states is the risk of malicious attack. Data at rest is at lowest risk of attack. Of course, data may be corrupted or destroyed on any storage device, so precautions should always be taken, such as backing up data.

Any time data is moved, hackers may intercept it, and this is particularly true when it is moved over public networks like the internet. Encrypting the data before it is transmitted (while at rest), encrypting the connection, or both will reduce this risk. Security is beyond the scope of this book, but the interested reader may consult [13]. When data is in use, it is directly accessible and at highest risk. The risk may be reduced by using encryption, user authentication, and permissioning.

4.4   Data Sources and Data Veracity

Data may originate internally at an organization, or it may be collected from external sources. Internal data might have been carefully collected with a survey, for example. External data can come from a wide array of sources. For any data set, internal or external, it is essential to understand how the data is collected and processed, and to ensure its quality. Reliable, useful data can enrich analysis and modeling. Bad data can tank an application and render the output useless.

Some datasets only need to be collected once, while others need to be collected on an ongoing basis. In the first case, a dataset might be downloaded for a research project or an academic project. In the second case, the data might need to feed an application running in production. An example might be daily stock prices that are ingested into a financial risk management application. It is essential in this case that the data arrives in an ongoing, timely manner.

Given the massive number of data sources on the internet, it can be challenging to find trusted data. Highly cited and used works will be preferable. To name a few sites for one-time download of datasets, the UCI Machine Learning Repository maintains over 500 freely available datasets, while Kaggle holds over 50,000 public datasets.

Here are some questions to ask when validating data sources, from [14]:

- What is the reputation of the source?

- What is the reputation of the author?

- Does anything jump out as being potentially untrue?

- Can this data be cross-referenced for accuracy?

4.5 Data Ingestion

Data ingestion is the process of collecting data from one or more sources for immediate use or storage in a centralized repository. The data may be ingested in real time (as it arrives) or in batches. The complexity of the data ingestion step can range from the download of one small file for an academic project to streaming video footage from a traffic camera.

When ongoing data ingestion is required, the job may be scheduled. For example, a system might collect daily batches of data from a vendor at 2 am Eastern Time. A utility program in the Linux operating system called `cron` allows users to input commands for scheduling tasks at a specific time. A task scheduled in `cron` is called a `cron job`. A task that runs at 2 am each day of the year can be scheduled with this command:

```
0 2 * * *
```

where

- The first value (0) indicates the minute

- The second value (2) indicates the hour

- The third value (*) indicates that the job runs every day of the month

- The fourth value (*) indicates that the job runs every month of the year

- The fifth value (*) indicates that the job runs every day of the week

4.5.1   Data Velocity and Volume

The *velocity* of data refers to the speed at which it enters a system and must be processed. Examples of high velocity data are electronic trades on a stock exchange and TikTok posts. Assuming the ongoing operation of each of these exchanges, the flow of this data is rapid. An example of low velocity data is the recording of lab measurements for a single patient during a one-week hospital stay.

We can think of data *volume* as the number of rows of data, such as the number of transactions, stock trades, or patient discharges. The volume of data in the stock exchange and TikTok examples can be treated as infinite. For the case of lab measurements, the volume is very low.

Let's now think about the relationship between volume and velocity, and how high levels of either can pose a challenge, using snowfall as an example. When I was living outside of Boston, one extreme winter brought so much snow, both in velocity and volume, that the city buses were picking up customers in the middle of the streets. The city needed to multiply their workforce to move the snow, and they were running out of places to put it. In this scenario, both processing (the number of workers) and storage (where to put the snow) was a challenge. Next, imagine a snowfall with high velocity, but lasting only a few hours. Not knowing the duration in advance, a scaled-up workforce might have gotten started carving out paths through bus stops and busy streets. In this beneficent scenario, the snow never accumulates, and there are plenty of places to move the snow. In these conditions, processing was a challenge but storage was plentiful. A third challenging scenario would be a slow but large accumulation over a longer period. In such a case, storage would be a bottleneck.

From these different snowy conditions, we learned that the only easy scenario is when both the volume and velocity of snowfall is low. Perhaps this explained my move to California after that winter! Bringing this back to the data, cases of high velocity, high volume, or both can cause challenges for a system in terms of storage and processing. In response to this data deluge and the growth of Big Data, new methods such as *map reduce* and new tools including Hadoop and Spark were born.

4.5.2   Batch versus Streaming

A finite block of data is called *batch data*, and processing the batch is called *batch processing*. Before sources like the web, mobile devices, and the internet of things, most data was batch data. A program can read in batch data from an input file, process the data, and store the results to an output file. After the results are stored, the job is finished, so things are straightforward.

Now think about a system that connects to X (previously Twitter), ingests all of the posts, extracts the tweets (messages), computes the frequency of each hashtag, and saves the results to an output file. This is an example of a *streaming system*, and processing this data is called *stream processing*. As the X feed is like a firehose, there is no end to the data ingestion, the extraction, the computation, and the storage of results. This causes some serious complexity such as:

- How can all of this data be stored? (Answer: It can't)

- How often should the computation be done? (Answer: This needs to be decided. For example, calculations might take place each time new data arrives.)

- How often should the results be updated in storage? (Answer: This again needs to be decided.)

There are many cloud-based services that streamline the process of establishing, securing, and automating data ingestion, including Apache Kafka, Apache Flume, and Amazon Kinesis. The important features include speed, security, connectors to data sources, automation, scalability, and ease of use. Next, we review approaches for ingesting data by leveraging code.

4.5.3 Web Scraping and APIs

Two common methods for ingesting large quantities of data are *web scraping* and the use of an API. Web scraping uses computer code to collect data from web pages found on the internet. When done properly, the process collects publicly available data following respectful conventions. For example, a user should not overwhelm a host website with a massive number of requests in a short period of time. Scraping can face several challenges, such as changes to the structure of the web page. Web scrapers work by exploiting the layout of a page, and if that layout changes, the scraper can fail to retrieve data. For a system running in production, a broken web scraper can be disastrous.

A popular Python module for web scraping is `Beautiful Soup`. As this is an introductory text, we will not go into the details of web scraping, but the short code snippet below illustrates how much can be accomplished with the module.

```
# Scraping with Beautiful Soup

# import packages: requests, BeautifulSoup
import requests
from bs4 import BeautifulSoup

# specify a website URL
url = 'https://en.wikipedia.org/wiki/
        Representational_state_transfer'
```

```
# send a GET request to fetch webpage data
response = requests.get(url)

# create BeautifulSoup object and input the webpage data
soup = BeautifulSoup(response.content, "html.parser")

# extract the title from the page
title = soup.find("title")

title
```

```
OUT:
<title>Representational state transfer - Wikipedia</title>
```

The **requests** and **Beautiful Soup** modules are imported for the necessary functionality. A **GET** request is sent to fetch the data from a Wikipedia page. Next, a **Beautiful Soup** object is created and it takes the page content as input. Lastly, the title of the page is extracted with the **find()** function. The output is shown below the OUT: line. The tokens **<title>** and **</title>** are HTML tags which indicate that the enclosed text is a title. HTML is a standard markup language for web pages, and elements of the page use various tags. These tags are used by **Beautiful Soup** and other scrapers to extract information of interest. We will learn more about modules, variables, and functions later in the book.

Next, we will have a brief overview of APIs. An API is a protocol for interacting with another system using an agreed-upon set of instructions. To help developers understand how to use a particular API, an *API Reference* document is commonly provided. This will include details on the supported methods, the expected parameters for making a request (e.g., passing an identifier for a desired model), the response delivery format and contents, and the meaning of various status codes, among other elements. The status code will alert the user to the outcome of the request. For example, a successful request may return status code 200, while an unauthorized request may return status code 401. A detailed understanding of the API will allow for efficient and proper data collection.

For security purposes, many APIs require users to provide credentials before requests, or calls, can be made. At the time of writing, a widely used standard for API authentication is *OAuth 2.0*. This provides a secure and convenient method for users to access resources without sharing passwords.

Most web APIs conform to the REST architectural style. Such APIs are called REST APIs, where REST stands for *representational state transfer*. REST follows guidelines to allow for scalable, secure, simple APIs:

- Client-server separation: the client makes a request, and the server responds to the request.

- Uniform interface: all requests and responses use the HTTP communication protocol. Server responses follow a consistent format.

- Stateless: the client-server interactions are independent. Each request contains sufficient data for successful completion.

- Cacheability: resources should be cacheable (stored for later retrieval and use) when possible. Responses must define themselves as either cacheable or non-cacheable.

- Layered system architecture: Requests and responses pass through a layered system architecture. One benefit of a layered system is scalability, or the ability to handle large workloads.

Many applications provide an API, and it is always advisable to check for one before attempting to scrape data. Some APIs are freely available to the public, while others are paid. For paid APIs, contracts specify terms such as how many requests can be made over a given period of time. Software packages exist to collect certain datasets with an API, such as the `wikipedia` package for Python. This package makes it easy to access and parse data from Wikipedia, as illustrated in the short code example below. For more details on REST APIs, see [15].

```
# Wikipedia API

import wikipedia

entity = 'New York (state)'

# use API to fetch webpage data
page = wikipedia.page(entity)

# extract the content from the page
page.content

OUT:
'New York, often called New York State, is a state
in the Northeastern United States. With 20.2 million residents,
it is the fourth-most populous state in the United States...'
```

The `wikipedia` module is imported for functionality. The API is used to get webpage data for the state of New York. The content is then extracted from the page with `page.content`. For more information about the module, including how to install it, see [16]

Scripting languages like Python and R facilitate data ingestion. The required coding is generally light, due to the availability of pre-existing modules.

4.6   Data Integration

Data integration is the process of preparing and combining the ingested data sources. As an example, consider a financial analyst that needs to determine if a company is likely to go bankrupt. There are many potential data sources that might be useful, including news, financial statements, historical stock price data, analyst reports, and videos from Bloomberg. After the various sources are ingested, they will need to be processed and combined. The video data can be transcribed to text, which may then be converted to numbers. The financial statements might hold tables of numbers, and these numbers might need to be stored in more useful structures. Data integration is the process that implements all of this work. To be effective at scale, it needs to be accurate, fast, secure, and automated.

Following the ingestion of data, it needs to be extracted, transformed, and loaded into storage, generally in this order. This sequence is so common that it is given the acronym ETL for Extract, Transform, Load. The object that implements the steps is often called the ETL pipeline. The extract step moves the raw, ingested data to a landing zone for staging. The landing zone provides temporary storage for the data integration and processing steps. After the data is finalized and moved to its destination, the landing zone is often deleted.

The transform step applies various functions to make the data more useful. This might include reshaping and filtering the data. The load step then stores the data into a repository, such as a data warehouse or data lake. ETL pipelines are the workhorses that implement these operations, and they facilitate the operability of systems. We will discuss data transformations and storage options in more detail later in the book.

An alternative to ETL is ELT. In this paradigm, the extracted data is stored first and processed later. This is common when the use cases and valuable data elements will be determined later. For example, data scientists might load the contents of scraped web pages, parse out elements of interest, and structure them into a table.

It can require a lot of effort to build and maintain the ETL pipeline, but there are many helpful tools. The growing list includes AWS Glue, IBM DataStage, and Oracle Data Integrator. Important features include speed, connectors to data sources, automation, scalability, and ease of use.

4.7   Levels of Data Processing

We reviewed the processes of data ingestion and data integration. This is very dynamic, with data moving within and between systems, data changing form, and data being created, modified, deleted, and stored. Data can also be categorized by its level of processing. *Raw data* is data that has not undergone any processing. The potential sources and types of raw data are very diverse and may include video camera footage, ECG waves, student survey results, and scanned documents. The benefit of capturing and storing raw data is that all of the data will be available, and the user can revert to the original state of the data as needed.

Consider the example of a photographer taking a picture of Times Square in the evening under difficult lighting conditions. The image is stored and later modified by changing the contrast and brightness, cropping the frame, and changing it to monochrome. The modified image is saved as a new file. The photographer later decides that she prefers the picture in color. Since the original photo is stored, she can revert back to it.

The limitation of raw data is that it likely won't be immediately useful for tasks like analysis, predictive modeling, or reporting. One reason is that raw data is commonly noisy. In the case of the ECG waves, the raw waves cannot accurately detect a heart arrhythmia due to artifacts such as patient movement and power line interference. Before the data can be trusted to answer questions, it needs processing. The specific processing steps will depend on how the data was collected, what defects might be present, how the data will be used, and who can view the data. In a proper system, the data processing will often be triggered automatically and implemented by computer code such as Python or Bash.

What Is Bash?
Bash is a Unix shell and command-line interface (CLI). It is a powerful, efficient tool for working with files. We will learn more about the shell and the CLI in Chapter 5: Computer Preliminaries.

Specific data processing steps may include:

- Segmenting the data into parts, such as breaking paragraphs into sentences, ECG waves into heartbeats, and comma-separated rows into columns

- Repairing foreign language characters which are transcribed incorrectly

- Verifying data by applying rules and logic, such as flagging rows of data containing an unexpected number of columns

- Detecting values that fall outside a predefined range

4.7.1 Trusted Zone

After the raw data is processed, the output data might be stored into a different zone on one or more computers. Different organizations might name the zones differently; we will call this the *trusted zone*. Concretely, there might be a folder containing the raw data files, and a separate folder containing the trusted data files.

An analyst using the trusted data files is less likely to encounter bad characters or other problematic data. However, errors are still possible, and one reason is inconsistency. For example, the definition of an outlier might change for a variable, and this might not be reflected properly in the data. In such a case, the analyst might need to revisit the raw data to rebuild the trusted data.

The trusted zone is typically not the last stop for the data. There may be processing steps that need to run on the trusted data, such as data standardization, natural language processing, and detecting and masking sensitive information. The steps should support the goals of the organization. A complication is that different users might require different versions of the data, as in the case of differing privacy requirements. Taking this into account, an organization needs to be strategic when deciding how to process, secure, store, and manage the data. If a certain step is required for all users, then it should take place once as part of the main process; it should not be done by multiple teams, as this can introduce inconsistencies and redundant work.

4.7.2   Standardizing Data

Data is often messy, and when it comes from multiple sources, it may be inconsistent. Suppose a system is collecting patient lab records from different laboratories. Even on a single lab test, there will be name variations. Here is an example of possible names for a COVID test:

- COVID-19

- COVID19

- COVID-19 (Lab X)

- Covid-19

In the Lab X case, a lab provider has included its name in the test name. These data variations happen in practice, and they make it difficult for systems and users to discern identical objects from different objects. This *entity resolution* problem is challenging and ubiquitous, and techniques such as machine learning can help.

This is just one example of standardizing data. It may be the case that every piece of data needs a standardization step. This is a lot of work, which is why automating these steps is so critical.

4.7.3   Natural Language Processing

A series of steps for processing language is often required to make the data more useful. This may include determining the dominant language of the text, splitting text into sentences and words, extracting keywords, classifying the document (e.g., is it a driver's license? Is it about sports?), detecting hate speech, and finding entities such as people, organizations, and locations. This is a very deep field with a lot of activity. See [17] for an excellent, detailed treatment.

4.7.4   Protecting Identity

Personally identifiable information (PII) and *protected health information* (PHI) such as phone numbers and social security numbers may be stored in a system. To preserve anonymity, this data often needs to be detected and masked. This is because some users may not have clearance to work with PII and PHI. One method for masking data is to replace sensitive information, such as a patient name, with random letters.

4.7.5   Refined Zone

After processing the trusted data, it is best practice to store the output into a separate zone. Here, I will call this the *refined zone*. Different organizations might name this data differently, divide it into several zones, or place it in different structures.

The data in the refined zone will likely be subjected to very specific treatment with different assumptions, algorithms, and so on. If all goes well, then this data should be very useful for solving problems. In the event of weaknesses, the appropriate users need to identify and improve them, backing up to the trusted zone and raw zone as necessary.

The table below summarizes the data zones by their degree of processing and reliability:

Zone	Processing	Reliability
Refined	Highest	Highest
Trusted	Moderate	Moderate
Raw	None	Lowest

4.8   The Structure of Data at Rest

When data is at rest, it may take one of three different forms, commonly referred to as *structured*, *semi-structured*, or *unstructured data*.

4.8.1   Structured Data

For some data, it makes sense to impose structure on it. The structure can bring benefits such as easier retrieval and downstream processing. The data may be stored in one or more tables with rows (also called *records* or *tuples*) and columns (also called *attributes* or *fields*). Records from a sample product table are shown below. The tables may be stored in a database to gain benefits such as control, efficiency, and security. An important step in designing the database is developing a schema, which serves as a blueprint detailing how the database will be constructed. We will learn more about databases later.

product_id	product_name	product_color	manuf_id
11112	Wham-O Hoola Hoop	Purple	1768
11113	Original Slinky	Silver	6587

A good candidate for this structuring is transactional data, as it often consists of common attributes. Product purchases on an e-commerce site might include dimensions, weight, price, color, and buyer details. A bad candidate would be the contents of unrelated web sites such as Yelp and NFL.com. While these are both web sites, the Yelp page might contain hours of operation, location address, and price, while the latter might include team records and game scores. Next, we look at less structured forms.

4.8.2   Semi-structured Data

Returning to the Yelp and NFL.com example, let's look at some sample data that might be of interest. From Yelp, we might be interested in the address and ratings of a restaurant.

200 E Main St
Charlottesville, VA 22902

number of 5-star reviews: 10
number of 4-star reviews: 5
number of 3-star reviews: 3
number of 2-star reviews: 1
number of 1-star reviews: 1

We might structure the Yelp data like this:

```
{'address' : {
            'street number': 200,
            'street name': 'E Main',
            'street type': 'St',
            'city': 'Charlottesville',
```

```
          'state': 'VA',
          'country': 'United States',
          'zipcode': 22902
     },
  'reviews' : {
                   'num_5_star' : 10,
                   'num_4_star' : 5,
                   'num_3_star' : 3,
                   'num_2_star' : 1,
                   'num_1_star' : 1
  }
}
```

Notice there is a hierarchy, where address and reviews are separate objects at the top level. Inside the address, there are components at the same level, such as street number and state. Here, 'street_number' is an example of a key, and it is associated with its value of 200. This is called a *key-value pair*. In this format, it would be straightforward for computer programs to retrieve data elements.

From the website NFL.com/standings, we might care about rankings in a division such as:

Division: AFC West
Chiefs
Chargers
Raiders
Broncos

Division: AFC East
Bills
Dolphins
Patriots
Jets

This data might be structured as follows:

```
{
    'division': {
        'AFC West':
            {'rank':['Chiefs','Chargers','Raiders','Broncos']},
        'AFC East':
            {'rank':'Bills','Dolphins','Patriots','Jets']}
    }
}
```

For example, the top-ranked team in the AFC West is the Chiefs. Forming these structures can be difficult by hand, but computers can facilitate the work. If we wanted one structure to hold both the Yelp data and the NFL data, we could add a unique key for each web page and merge these two structures together.

This semi-structured format, called JavaScript Object Notation and shortened to JSON, is extremely popular due to its ease of use and flexibility. In fact, it is the universal standard of data exchange through APIs. In other words, when applications send data to another system, or retrieve data from a system, they use this format. For each supported method, a JSON object will be used to make a request, and another JSON object will return the response.

An excellent website for testing and editing JSON is:

`https://jsoneditoronline.org/`

As we will see later, Python has a data type called a *dictionary* which supports key-value pairs in a fashion similar to JSON.

4.8.3   Unstructured Data

E-books, webpages, video transcripts, and clinical notes are just some examples of sources containing rich information that cannot be easily entered into a tabular form. As a result, collection, processing, and analysis of this unstructured data is more challenging than structured data. Imagine a publisher wishing to extract keywords and a topic list from tens of thousands of videos. This is a challenging task that can be done, but it will require transcription and machine learning.

When unstructured data is ingested, it is often directly stored in files without modification. This is particularly true for high velocity, high volume unstructured data. The common practice is to store this data in a data lake, which will be discussed later. If specific information is needed from the data, then it is good practice to design a schema, extract the data, and store it in a database for easier retrieval.

4.9   Metadata

Metadata is data about data. Metadata about a book would include the title, publisher, author, ISBN, and table of contents. Movie metadata such as runtime, producer, genre, and actors tell us about the movie, and perhaps whether it would be of interest. Particularly, as we are awash in more data than we can possibly review, metadata is useful for finding things, making recommendations, and understanding the context and lineage of data. Data can be made more useful by extracting and storing its metadata.

4.10 Representativeness and Bias

When we think about an object or event (subsequently called *things*), whether it's a deck of cards, a flower, or a shuttle launch, we reduce it to some attributes to generate understanding. We might identify the deck of cards as such because there are 52 pieces of identically shaped paper of certain dimensions, marked with a familiar set of colors, numbers, and patterns. But what if the cards are made of wood? What if there is an extra queen of hearts, or a missing queen?

We understand things by reasoning about their data, and in the other direction, we can reason about data and attempt to reconstruct things. This works well when the data representation is sufficient, but when essential information is missing, there is *bias*. A goal of useful data is that it should be *representative* of the thing it seeks to represent.

Bias is a major subject of statistics, and one reason is that sampling methods will infer properties from a population of interest. If the data is biased, it will be systematically misrepresentative of the overall population. When bias is combined with automation, the consequences can be disastrous.

Here are some examples where bias can have harmful consequences:

- A school district conducts an internet survey of parents to understand if they need subsidies to afford lunch. The district is located in a rural town, where many residents don't have internet access.

  This is an a example of *selection bias*, whereby the sample of individuals is not representative of the population. The risk is that the survey results may not address the needs of parents.

- A county would like to raise additional funding by opening a large shopping mall. The mall would be placed on a sensitive Native American burial ground. To understand how the residents feel about this project, a random sample is drawn and asked to participate in the survey. The fraction of Native Americans in the county is less than 1% and it happens that none of them are invited to participate.

  In this case, the minority group is very small and happens to be excluded from the survey. This results in *sampling error* since the Native Americans are not represented.

The challenge with bias is that it can easily happen unintentionally, and it can result from a technical shortcoming. To help avoid bias, ask:

- Are we leaving out important time periods?

- Are we leaving out important groups, such as demographic segments?

- Are we leaving out important locations?

- Do the group proportions in the sample match the proportions in the overall population?

Finally, all people working with data should ask these questions – not just data scientists.

4.11   Data Is Never Neutral

As we learned from the bias discussion, data is never neutral. How the data is conceived, collected, processed, stored, analyzed, visualized, and used will differ by person. This means that as we work with data, we must learn to view each step critically. We should ask questions like:

- Where did this data come from, and how was it created?

- How might this data be biased?

- What are the limitations of this data?

- What can we do with this data, and what can we not do?

- How can we improve this data?

- Is this analysis interpreting the data correctly?

Challenge the data, the analysis, and the results. The application of critical thinking to data and the products of data will always be valuable and important.

4.12   Chapter Summary

Data is information that can be stored and used by a computer. Each piece of data has an associated data type, such as *string*, and a statistical data type, such as *categorical*. Data types define the values and operations that are possible for the data, such as whether the values can be sorted or added.

Data ingestion is the process of collecting data from one or more sources. The data may be used immediately or stored for later use. Two important approaches for data ingestion are use of an API and web scraping. The API delivers data in an agreed-upon format. Web scraping can be an effective method for collecting data, but it can encounter many challenges such as changes to the structure of web pages.

Data integration consists of preparing and combining the ingested data. The common pipeline consists of an extraction step, a transform step, and a load step. This is called the ETL pipeline. In some cases, such as when the important data elements are not known, the data is stored without processing.

Raw data is unprocessed data. As data is increasingly cleaned and processed, it becomes more trusted for downstream applications. Metadata is data about data, such as the genre and runtime of a movie. This information can be helpful when searching for relevant objects.

Data can be at rest, in use, or in motion. The velocity of the data is the speed at which it enters a system for processing. We will measure data volume as the number of rows of data. High-velocity and high-volume data necessitate specialized tools for storage and processing.

Some datasets are finite and they can be processed in a large chunk. This is called batch processing. Data that is infinite, such as video from a sensor, is called streaming data. This kind of data requires decisions such as what data to store (since it can't all be stored) and when to compute (since there is no end to the data).

Data at rest may be structured, semi-structured, or unstructured. Structured data can be stored in tables, and transactions are an example. Semi-structured data is generally hierarchical, and it is stored as key-value pairs. Home address is an example of semi-structured data. Unstructured data comes from a variety of sources, such as audio files and images. This format cannot be easily stored into a tabular form.

To be useful, data needs to be credible and reliable, and it must faithfully represent the population. In cases where the data is not representative, there will be bias. Care needs to be taken to prevent and mitigate bias, as it can have harmful consequences and lead to poor decisions.

In the next chapter, we will briefly study computing preliminaries that every effective data scientist needs to know. This includes hardware, software, and version control. We will also get started with essential tools including GitHub.

4.13 Exercises

Exercises with solutions are marked with Ⓢ. Solutions can be found in the `book_materials` folder of the course repo at this location:

https://github.com/PredictioNN/intro_data_science_course/

1. Ⓢ Which of the following are data? Select all that apply.

   a) a ruler
   b) the dimensions of a ruler
   c) a book
   d) the text from a book
   e) daily historical temperatures

2. An alternative to the REST API is the Simple Object Access Protocol, or SOAP, API. Explain the major differences between the two. Why is the REST API generally preferred?

3. Data ingestion with an API is preferable to web scraping. Explain why this is the case.

4. Explain the difference between raw data and trusted data.

5. Ⓢ A set of patient ECG waves are collected and processed to remove noisy measurements. The cleaned data is stored and the raw data is deleted. Is this a good practice? Explain why or why not.

6. Ⓢ For each of the given examples, indicate if the data is at rest (R), in use (U), or in motion (M):

   a) restaurant reviews on Yelp
   b) a file in the process of being downloaded to a laptop
   c) a table of data being updated by a nurse
   d) a set of videos being transferred from a camera to a server

7. Ⓢ Imagine that you have developed a system that reports on the risk of clients' stock portfolios. You have an automated, daily data feed that delivers data to your system at 1 am Eastern Standard Time. Occasionally, some stock prices arrive up to 24 hours late. Clients expect a statistical report each day. Explain how you might calculate and report the statistics to satisfy your clients. Results don't need to be perfect.

8. Ⓢ You browse Netflix to find a movie to watch with your friends. How can metadata help in making the right selection?

9. Provide an example of semi-structured data that you might encounter in real life.

10. Explain selection bias in simple terms.

11. Ⓢ What is the danger of bias in a dataset?

12. Ⓢ Imagine that data flows through a pipeline with these steps:

    Step 1: Sensors capture the data
    Step 2: Algorithms process the data

Step 3: Predictions are made by a machine learning model
Step 4: Results are reported in a dashboard on the web

Given that each of these steps is automated, is it correct to say that
the data is neutral? Explain your reasoning.

5

Computing Preliminaries and Setup

We have covered a lot of information about data and data science project planning. Doing data science requires computing, and this section will cover the basics for getting started. Machine learning models are pushing computational limits, and this requires the latest hardware. We will discuss the essential hardware components of a computer.

Many new data scientists struggle with operating system differences, file paths, input/output, and the *terminal*. While most corporations run the Linux operating system, many new data scientists have never used Linux. The terminal can be intimidating to the uninitiated, as it is a blank screen. We will dive in and get experience running commands in the terminal.

The term "cloud" has been briefly mentioned earlier. This chapter will outline *cloud computing*, the major players in this space, and the services that they offer. Companies seek data scientists with experience in the cloud, so this is an important skill.

Data science is done in teams, and collaborating on code is a daily necessity. We will discuss how to use `Git` for version control and `GitHub` for collaborating on coding projects. As a side benefit, this will provide the background for working with the code repo for this book.

Lastly, we will cover some fundamentals of computing with Python. This will include different ways of running Python code, and using pre-existing code. Following this chapter, there will be many opportunities to practice and strengthen Python skills.

5.1 Hardware

Hardware consists of the physical components of the computer. We will focus on the main components that help us do data science: processors, RAM, hard drive, and the motherboard.

5.1.1 Processor

The *central processing unit* (CPU) is the brain of the computer. It is constructed from millions of transistors, and consists of one or more cores. Each core receives instructions from a single computing task, processes the information, and temporarily stores it in the *Random Access Memory* (RAM). The results are then sent to the relevant components [18].

The CPU is critical in determining how fast programs can run. Major factors that determine the processing speed are the number of *processor cores*, the *clock speed*, and the *word size*. The clock speed is the number of cycles the CPU executes per second, measured in gigahertz (GHz). The word size is the amount of data that can be transferred to the CPU in one pass. A good clock speed for today's machines is 4 GHz, which means 4 billion calculations per second. A typical word size is 32 bits or 64 bits, where a bit is the smallest unit of data stored by a computer. Each bit will take a value of 0 or 1. For both the clock speed and the word size, a larger value will mean faster processing.

The *graphical processing unit* (GPU) is a processor made of specialized cores. Like the CPU, it is a critical computing engine. However, its architecture is different, as its purpose is to accelerate 3D rendering tasks [19]. It does this by performing parallel computations, which happens to be very useful for heavy machine learning workloads. Today, some GPUs are designed specifically for machine learning, and some machine learning algorithms and frameworks – specifically deep learning, which is a subarea of machine learning – are designed for GPUs.

The technology of CPUs and GPUs has been rapidly improving, which has supported major advances in data science. Today, GPUs may be integrated into CPUs. Resource intensive tasks, such as computer vision, benefit from the combination of CPUs and GPUs. It is common for a data scientist to think about the task at hand, and select a machine with a desirable number of CPUs and GPUs.

5.1.2   Memory (RAM)

Random Access Memory (RAM), generally called memory, provides a place to store data for the CPU. At any given time, the operating system is often running several applications such as a browser, an email application, and a design tool. Memory provides short-term storage to quickly move data in and out of the CPU.

RAM is much faster than the hard drive storage we will review next, but it is much more expensive. As a result, a computer will come equipped with much less RAM than hard drive space. A typical laptop will have 8 GB or 16 GB of RAM (32 GB for a higher-end machine), while the hard drive space might be 500 GB to 1 TB (=1000 GB).

When doing data science, 16 GB of RAM will be enough to store relatively small datasets for processing. Breaking the data into smaller pieces, or *chunking*, can allow for handling larger datasets. However, sometimes a dataset needs to be processed and it doesn't fit in RAM. In this case, another option is required, and this is where cloud computing can be helpful. Cloud computing allows access to additional compute resources, including more RAM,

CPU, and GPU. We will discuss cloud computing in more detail later in this chapter.

5.1.3 Storage

The computer hard drive, also called disk space or *disk* is another option for data storage. Disk is used to store files supporting the operating system, files for running applications, and other miscellaneous files. Disk is much less expensive than RAM, but it is also much slower for file access, which translates to slower processing speeds.

The two main types of disk storage are the older *hard disk drive* (HDD), technology and the newer *solid-state drive* (SSD). The HDD consists of a spinning platter (the disk) and an arm containing several heads. The arm moves across the platter to read and write data. SSDs are mechanically very different, as they store the data in integrated circuits. As they don't use a spinning disk and arm, they can be smaller, and the time to read and write is much faster. SSDs are generally more expensive than HDDs, but the cost has been falling over time.

5.1.4 Motherboard

The *motherboard* is a board with integrated circuitry that connects all of the principal hardware components. It distributes electricity from the power supply to the essential components, including the CPU, GPU, RAM, and disk.

5.2 Software

Software is the set of instructions used to operate computers and execute tasks. It can be stored on disk or in memory. The most important software on the computer is the *operating system*, or OS. The OS manages the computer's memory and processes, including all of its hardware and software. Bringing up the system information page will show the OS name, version, and other specifications.

There will be some user experience differences depending on the running OS. Visual differences will be most apparent, as the backgrounds, icons, and window panes will look different. Functionality will also vary between the operating systems, and next we will discuss how directories are structured.

Tip: The Mac operating system derives from the Linux operating system, and this explains why some of their functionality is the same or very similar. This also explains why instructions may be grouped for Mac and Linux users. This book will use this convention as well, treating Mac as a catchall for the Mac and Linux operating systems.

5.2.1   Modules

Programming languages provide the capability to save and distribute software with modules (also called *libraries* or *packages*). For commonly used functionality, such as matrix computations, data processing, and statistical modeling, this can help others save tremendous amounts of time and effort.

In particular, Python users can leverage *Python modules*, which organize functions and structures into files. We briefly saw examples which used the `requests` and `Beautiful Soup` modules. Anaconda has many essential modules pre-installed, and it is straightforward to install many others by running the command `pip install [package_name]`, where `[package_name]` is the package name. This may be run in a terminal or a Jupyter notebook, for example. We will review these applications later in this chapter.

To use an installed module, the `import` statement needs to be run. It is possible to import the entire module or only the required parts, such as specific functions. For example, the `os` module for operating system functionality can be imported by running this command:

```
import os
```

If a user wants to use a specific function from the `os` module, such as `chdir()`, this can be done by running

```
os.chdir()
```

We will import many modules throughout this text. It will become very evident how we can build on the impressive work of others.

5.3   I/O

Input/output (I/O) refers to the communication between information processing systems. In this book, it will be most common to read data from a file, which is the input, and use Python code as a set of processing instructions for the computer. When the computer is finished processing the data, it will send the result as output. The output might be a new file containing data, a beautiful graph, or some other form.

5.3.1   Directories and Paths

Operating systems will organize files into a directory structure. The directories consist of folders, and the folders may contain additional folders; this is called nesting. Files are saved in these folders. For specifying where input can be found and where output should be saved, it is necessary to provide file locations, or paths. New data scientists may be surprised to learn that paths look different across operating systems. For example, a file path on Windows might look like:

`C:\Users\user_name\projectx\datasets\file.txt`

while on a Mac it might look like:

`/home/user_name/projectx/datasets/file.txt`

Pathing can be a major source of confusion, so this section will outline how it works. There are two differences in the appearance of paths: the drive letter and the slashes. Windows operating systems use a directory for each drive. Here, datafile.csv is stored on C:, which is the hard drive. Mac operating systems, by contrast, store everything in a single directory.

Secondly, Windows machines use backslashes "\" in paths, while Macs use forward slashes "/". When coding in Python (on any operating system), the forward slash is required for paths.

> Tip: Python uses backslashes as an escape character, which is a special character that invokes an alternative meaning of the characters that follow.

There are two options when specifying file paths. The *absolute path* states the entire path, while the *relative path* provides the location from the *working directory*. The working directory is where the system would start searching for files.

Our current example uses an absolute path. A quick indicator is that a drive letter appears at the start of the path. There are advantages to using relative paths. The first is that the absolute path can get long if a file is deeply nested in the hierarchy. The second, which is extremely important, is that relative paths make code *portable*. Absolute paths are generally not portable, because different machines will generally use different paths.

For example, consider a code snippet with an absolute path to a home directory like `/home/apt4c/projectx/datasets/file.txt`. If a different user ran this code on a different machine, it would break because the file would not be found at that path. Instead, a relative path can be specified in the code. Suppose that the working directory is `/home/apt4c/projectx/`. This would be where the system would begin a search for files. We provide the path from that location, including the file name, which looks like this:

`'./datasets/file.txt'`

The dot denotes the current location and `/datasets/` directs the search into the `datasets` folder for the file. If a different user runs this code from the working directory `projectx`, then the relative path would be successful. This avoids the issue of machines having different paths earlier in the directory. To summarize this important point:

For portability, always use relative paths in code.

Later in the book, we will need to use relative paths that back up one directory. For example, this would mean navigating from

```
/home/user_name/projectx/datasets/
```

to

```
/home/user_name/projectx/
```

The Linux command .. will back up one directory. The command ../.. will back up two directories, and this pattern can be extended.

5.3.2   File Formats

The file format is the structure of data in a file. The format can make processing easier, as it can follow conventions such as separating columns by a designated character, or compressing data strategically. There is a wide range of formats for specific kinds of data, such as MP4 for videos, JPG for images, and CSV, TXT, or DOCX for text files.

The formats we will use the most for data are CSV and TXT. The CSV format separates columns with the comma delimiter. It can be very useful when each row contains the same number of columns. Here is a small example of what the first few rows of a CSV file might look like:

```
year,location_captured,females,males,cubs,unknown_sex,total
2014,China,14,11,2,0,27
2014,Russia,25,21,3,2,51
```

This dataset includes a header row which provides the names for each field. The header is not always included, and it is best to check this before plunging ahead into analysis.

The TXT format is very useful when the data doesn't follow a structure, such as the contents of a web page or a document.

A format that is increasingly useful in big data is the *Parquet format*, which stores a dataset by its columns, along with its metadata. It provides efficient data compression for fast storage and retrieval.

Python provides functions for easily saving and loading file formats. It won't be necessary to spend time writing code to loop over rows of data and load them in memory. This means there will be more time for doing data science and less time spent parsing data.

When we call a Python function to load data, the dataset is saved in RAM. It is important to verify that there is enough RAM to store the data in computer memory. For example, a machine with 16 GB of memory will not be able to load a 100 GB video file; such an attempt will fail with an *out of memory exception.*

5.4 Shell, Terminal, and Command Line

The ability to interact with the operating system of a computer is provided by the *shell*. The shell is a computer program that takes commands, interprets them for the OS, and outputs the results. It is called a shell because it wraps around the operating system.

Most Linux and Mac operating systems use Bash as the default shell. Windows machines come preloaded with two shells: `cmd` and `PowerShell`. There are several other shell programs, but at a basic level, they all support user interaction, and they differ in some of their functionality. Some power users can get quite particular about their shells!

There are two ways that users can interact with the shell: visually through the *graphical user interface* (GUI), and through the CLI. Most users are familiar with clicking icons in a GUI. Use of the CLI is often through a terminal environment.

On a Mac, users can bring up terminal with Spotlight (search for "terminal"). On a Windows machine, clicking the windows button and typing "terminal" will bring up PowerShell.

On a Mac, the terminal icon looks like the black square shown below.

Terminal icon

The > symbol is called the prompt, and it is where users enter commands. The line where commands are entered is called (very sensibly) the command line.

Launching Windows PowerShell brings up a black window with a command line consisting of a prompt. The window and the command prompt are shown below.

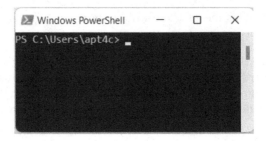

Command Prompt:
C:\Users\apt4c>

The prompt displays the working directory followed by a blinking cursor. This is usually intimidating for the uninitiated, as there is not much happening. It is waiting for a command. The contents of the working directory can be listed with the ls command, followed by Enter.

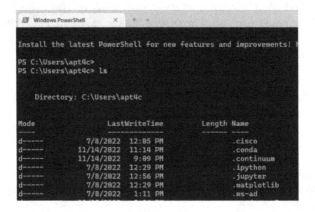

The terms shell, terminal, and command line tend to be used interchangeably, and this is because they are fundamentally intertwined. The shell is the program offering interaction with the OS, the terminal is a specific environment for the interaction, and the command line is the actual line where commands are entered. The book will use statements such as "open a terminal" or "run this at the command line," and clarification will be given as needed.

The command line offers two major benefits over the GUI. Since all of the commands are typed with the keyboard, this becomes extremely efficient as the commands and keyboard shortcuts are memorized. This avoids repeated movement between the mouse and the keyboard. Additionally, since all commands are typed, they can be easily shared with others trying to replicate the work. The commands entered in the terminal are stored, and they can be pulled up with the `history` command. By comparison, replicating work in the GUI requires screenshots or other media to clarify the steps; this creates additional work for everyone.

There is a lot more that can be said about the command line, and we will cover what we need during our journey. For an excellent, comprehensive treatment of command line tools, see [20].

5.5 Version Control

Version control is an essential tool for tracking file changes. For software projects, this is a must, as version control supports abilities such as viewing previous file versions, easily finding changes that a contributor has made, and ensuring everyone on a project is working on the same version.

5.5.1 Git

The most widely used modern version control system in the world is `Git`. It is mature, open source, optimized for performance, and secure. To track history, it focuses on changes in the file content, or *diffs*. This means that Git won't be fooled by changes to a file's name; only the file content matters for tracking diffs.

`Git` is a very powerful tool that provides a set of commands that can be entered in a terminal or in a GUI. For details, see [21]; the book is freely available here: `https://git-scm.com/book/en/v2`.

The first step is to install `Git`, which can be downloaded from here:

`https://git-scm.com/downloads`

> Tip: On a Mac, it may be easier to type `git` in the terminal and install the command line developer tools. These tools include `Git`.

We will use the terminal in this book when working with `Git`. The following illustration will use PowerShell. After typing `git` and pressing `Enter`, a set of commands will be displayed, as shown below. If these commands aren't listed, `Git` did not install properly. Try reinstalling `Git` in this case.

It will be valuable to learn the range of `Git` functionality. Data scientists typically use `Git` as part of their daily workflow, and with practice comes expertise. The optional appendix at the end of this chapter discusses more things that can be done with `Git`. These tasks will not be necessary in the scope of the book.

5.5.2   GitHub

Next, we will review `GitHub`, which is a popular internet hosting service for software development and version control built on `Git`. `GitHub` also provides tools for software feature requests, task management, continuous integration, and project wikis. For more details on `GitHub`, see [22].

`GitHub` and `Git` are often confused, and clarification is in order. `Git` can be used for version control independently of `GitHub` (or similar services like `GitLab`). However, tools like `GitHub` are required to collaborate on version-controlled software over the web with other users. Many users opt for `GitHub` even for themselves, as it provides a nice user interface and many helpful features.

5.5.3   GitHub Setup and Course Repo Download

In this section, you will create a `GitHub` account, make an independent copy of the course repo in your account (*fork* the repo), and download the repo to your computer (*clone* the *forked copy*). This will provide ongoing access to all of the materials. Figure 5.1 illustrates the repo objects. When a repo is cloned, a link is created between the *local* copy on the computer and the *remote* copy in the `GitHub` account. The arrow shows that link.

Creating a GitHub Account

If you don't have a `GitHub` account, visit the `GitHub` site to create one: https://github.com/

The signup process will ask for an email address, username, and password.

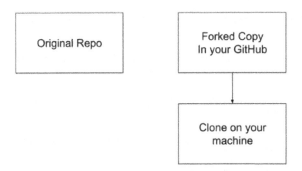

FIGURE 5.1: Original Repo, Forked Copy, and Clone

Next, create a personal access token, which is an alternative to using a password for authentication. Follow these steps[1]:

1. In the upper-right corner, click your profile photo, then click Settings

2. In the left sidebar, click Developer settings at the bottom, then click Personal access tokens

3. Click Tokens (classic), then generate new token (classic)

4. In the Note field, enter a descriptor for the token, such as main_token

5. Enter an expiration date

6. Click the repo button at the top, and scroll to the bottom

7. Click Generate token

8. Copy and paste the token to a file and keep it secure

Now you're ready to fork the course repo.

Fork the Repo to Your GitHub Account

Open another tab and go to the GitHub course repository found here:

`https://github.com/PredictioNN/intro_data_science_course/`

The course repo landing page, shown in Figure 5.2, contains folders for each semester, instructions, and a `README` file which serves as the syllabus, among other things.

To fork the repo, click the Fork button at top right.

---

[1]Note that over time, these steps may change. You can search for `GitHub` documentation on creating a personal access token for the latest steps.

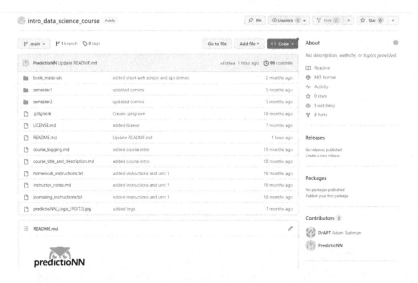

FIGURE 5.2: Course Repo Landing Page

Next, scroll down and click `Create Fork`.

Then visit your personal `GitHub` page. The repo should appear in your account.

Tip: Users can make changes to the forked copy, such as adding helpful notes and files. This will not affect the original course repo. Additionally, users can fetch updates from the original course repo and submit changes for review with a pull request.

Clone the Forked Copy to Your Computer

In this section, you will clone the repo from your `GitHub` account to your computer. To clone the repo, click the green Code button.

Next, with the HTTPS tab selected, click the copy icon, which will copy the URL of the repo. Note that in place of `PredictioNN` in the box shown in Figure 5.3, you will see your username.

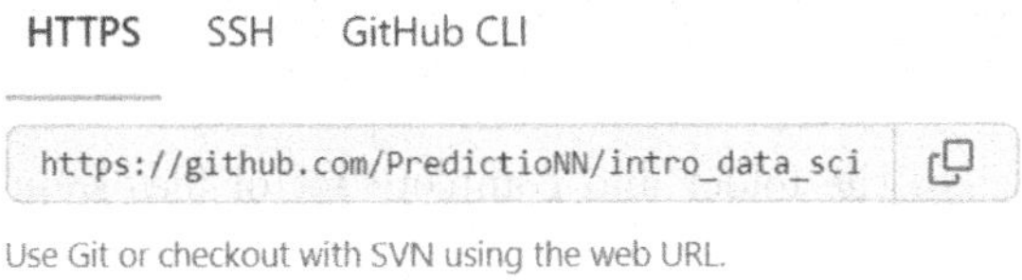

HTTPS   SSH   GitHub CLI

```
https://github.com/PredictioNN/intro_data_sci
```

Use Git or checkout with SVN using the web URL.

FIGURE 5.3: Clone the Repo

Next, open a terminal and change the directory (using `cd` command) to a desired destination for the course repo folder.

Figure 5.4 illustrates changing the directory to

`C:\Users\apt4c\Documents\repos`

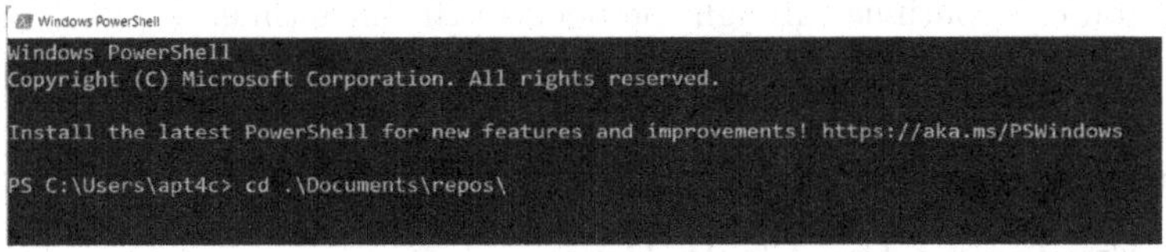

FIGURE 5.4: Clone the Repo

Next, clone the repo with this command:

```
> git clone [paste the copied repo URL here]
```

For this example, the command is:

```
> git clone http://github.com/PredictioNN
/intro_data_science_course.git
```

Figure 5.5 shows the directory change and `clone` command. Notice how the prompt updates to the changed directory.

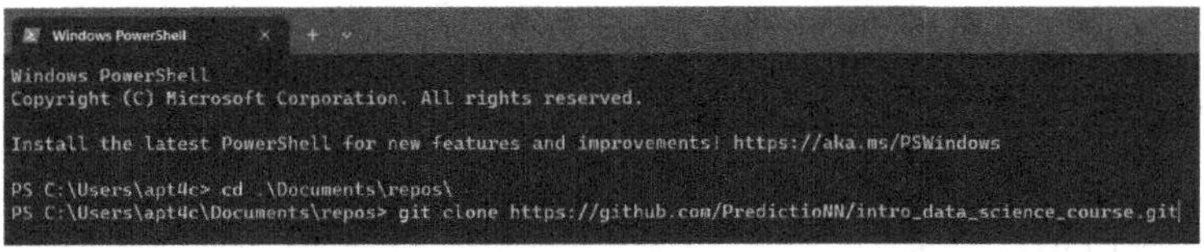

FIGURE 5.5: Clone the Repo

Running `git clone` will prompt the user for a username and password. The personal access token can be used for the password.

Once authenticated, the folder and complete set of files should copy to the specified path. Figure 5.6 shows the input and output for this example. If this succeeds, give yourself a big congratulations!

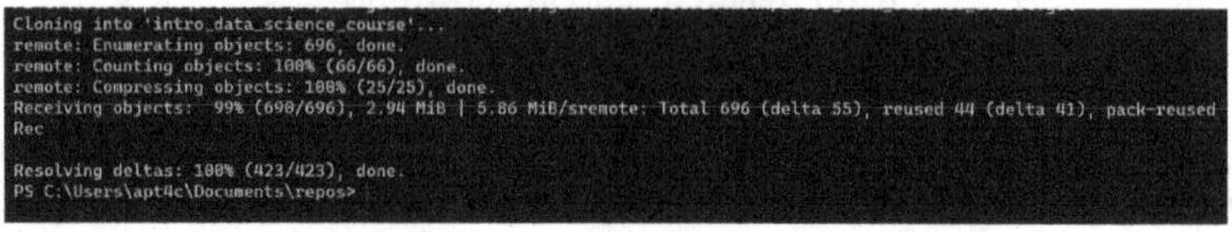

FIGURE 5.6: Clone Process Logging

If you ran into trouble or found any of these steps confusing, take your time, visit the `Git` and `GitHub` help pages on the web, and return when you are ready. You can continue through the book in the meantime.

5.6   Exploring the Code Repo

We take our first steps with the code repo in this section, learning about some of the applications of data science. If you were able to complete the last section, then you will have the repo in your `GitHub` account (the forked copy) and also on your computer (the cloned copy). In `GitHub`, let's visit the folder

`semester1/week_01_intro`

It is possible to click on the folders to navigate, and the path should be displayed like this:

intro_data_science_course / semester1 / **week_01_intro** /   ⊡

| Tip: You can click into this path to navigate backward as needed. |

The file we want to browse is called `what_kinds_of_problems_can_ds_address.pdf`.
Click into this file and take a look at the examples of interesting data science applications. Next, you will view the pdf file on your computer (the cloned copy). Open File Explorer (on Windows) or a Finder window (on Mac or Linux) and navigate to

`semester1/week_01_intro`

If you have an application to view pdf files, like Acrobat Reader, you should be able to open the file.

If that worked, congrats! You are now able to work directly in the course repo.

Notice there is another version of this file with the name:

`what_kinds_of_problems_can_ds_address.ipynb`.

This file can be directly viewed in `GitHub`, and it contains the same information as the pdf version. The `.ipynb` extension signifies that it is a Jupyter notebook file. In the next section, you will install `Anaconda` on your machine. This provides support for running the notebooks, among many other things.

5.7 Coding Tools

`Anaconda` provides an easy way to get started with Python. It's free, open source, and it installs with many common modules for productive data science.

Visit this site to download the installer:

`https://www.anaconda.com/products/distribution`

It should detect your operating system and figure out the right software version. Here is an example recommendation of the Windows 64-Bit version (Figure 5.7). Currently, Python is on version 3.9.

FIGURE 5.7: Anaconda Recommended Distribution

Next, step through the install process and follow the recommendations.

After the install is finished, open the Anaconda Navigator, which provides many fully featured tools.

We will use `JupyterLab` to run the notebooks, so launch that application.

`JupyterLab` will open in a browser. A file explorer window will appear in the left pane, and a file workspace will appear in the right pane (Figure 5.8). Navigation to a different folder can be done in the explorer. For the illustration below, I navigate to the course repo and open the notebook:

`what_kinds_of_problems_can_ds_address.ipynb`

The file opens, and we see that the content is arranged in cell blocks.

FIGURE 5.8: JupyterLab Session

The cells can contain *Markdown* (rich text with a specific format) or Code. It is possible to toggle between Markdown and Code using the drop-down shown here:

There are menus that support working with files and changing settings, among other things. Keyboard shortcuts are supported and allow actions like changing formats, adding cells, deleting cells, and running them. It is very helpful to learn the shortcuts to move fast, but they are not critical.

Inside a cell, it is possible to run the cell by pressing `Shift + Enter`.

`JupyterLab` automatically saves the notebook every two minutes; this is called *checkpointing*. Users can prompt it to save by clicking the save button (the floppy disk icon left of the Markdown drop-down).

The section at the top right displays a notebook kernel such as `Python 3` `(ipykernel)`. The notebook kernel is a computational engine that executes the code in the notebook. For the purpose of this book, it won't need to be changed, but there are a few things worth knowing about the kernel at this point:

(1) Sometimes the kernel dies (this is typically rare), or a computation stalls

or runs for a long time. The kernel can be restarted by selecting `Kernel >
Restart Kernel` from the main menu at the top.

(2) Different kernels can be prepared and loaded for different purposes. For
example, a user might have a kernel that runs a deep learning framework.

Tip: Although JupyterLab is running in a browser, it is actually running
locally on your machine and not the internet. In fact, an internet connection
is not necessary to run the application and notebooks.

If you made it to this step and you can run a Jupyter Notebook in `JupyterLab`,
I commend you! You have completed all of the setup needed for the book.

5.7.1 IDEs

Interactive Development Environments, or IDEs, provide a rich environment
for developing, debugging, and testing code. They include several panes for
various functionality such as writing code, browsing files, and interactively
running code.

Many IDEs have advanced, time-saving features including code completion, text highlighting, and variable exploration. Anaconda comes with an IDE
called *Spyder* which supports Python. Figure 5.9 shows the default multi-pane
display of Spyder, with a notepad for file editing at left, an object browser
at top right, and an interactive Python console at bottom right. Users can
run code at a command line in this console. It is also possible to access the
command history from a separate tab.

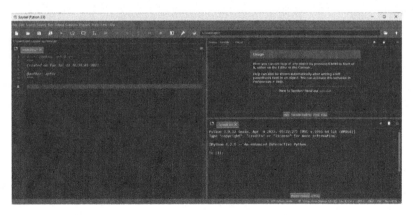

FIGURE 5.9: Spyder IDE

Another currently popular IDE is called *PyCharm*. It has an abundance
of useful features, including support for multiple languages and frameworks.
There is a free community version and a paid professional version. For more
details, see `https://www.jetbrains.com/pycharm/`.

While the IDE is very useful for software development, we will use JupyterLab in this book. JupyterLab is very useful for interactive scripting and demos.

5.8   Cloud Computing

Cloud computing is the delivery of computing services over the internet. It provides tremendous flexibility compared to computing with the machines *on premises*. Advantages include:

- Access to a wide variety of processors and storage, including state-of-the-art GPUs

- Automated scaling of hardware to match demand. This can be very cost effective compared to buying machines to match peak demand and having them sit idle for the remainder of the year.

- Highly durable data storage that reduces the risk of data loss

- Features for automating the deployment of services, such as model predictions

- Advanced security features

Cloud service offerings are commonly grouped into categories which often interact:

- IaaS, or *infrastructure as a service*, provides access to servers, storage, and networking. This is the back-end infrastructure required for running applications and workloads.

- PaaS, or *platform as a service*, is a cloud environment that includes everything needed to build, run, and manage applications

- SaaS, or *software as a service*, is on-demand access to cloud-hosted application software

Cloud providers own and maintain massive warehouses of hardware, and customers can access these resources virtually through an internet connection. Security is essential in this setup, whereby an organization's data is transmitted and placed on external servers which are shared by others. Customers pay only for what they use, which is called *consumption-based pricing*.[2] Providers offer a variety of hardware and software for different use cases and patterns,

---

[2]Note that forgetting to shut down a server, an issue called "leaving the lights on," will incur billing charges.

such as compute-intensive jobs and storage-intensive jobs. The warehouses are secure and climate controlled, and they are placed throughout the world to offer fast response times with high availability. Data can be copied to multiple locations to reduce the likelihood of data loss. This practice promotes *durability* of data; note that each copy of data incurs charges. Compute jobs can be set to automatically scale as the required resources increase or decrease. For example, additional servers can be added if there is a surge in customer demand.

The number of available cloud-provider services has grown rapidly in the last ten years. Early services included compute (for processing a batch of data with Python, for example), object file storage, and database storage. Today, there are services for video game development, end-to-end machine learning, generative AI, chatbots, and hundreds of other applications. The services respond to customer demand, and competition works to guide prices lower.

The largest cloud providers currently are Amazon Web Services (AWS), Microsoft Azure, Google Cloud Platform (GCP), Alibaba Cloud, and Oracle Cloud. AWS was first to market and it is still the largest cloud provider by market share. The top three players are fierce competitors offering similar services. Some of the key differentiators are the number of computing regions/zones, the pace of innovation (e.g., better hardware, fully featured services), the level of customer service, and pricing. For example, some providers build their own GPU chips to accelerate machine learning and bring down pricing.

The table below lists example services offered by the three largest cloud providers. Data lakes and relational (SQL) databases will be discussed later; the former is used for unstructured data, while the latter is used for structured data.

Service	AWS	Azure	GCP
Compute	Amazon EC2	Virtual Machines	Compute Engine
Data Lake	Amazon S3	Blob Service	BigQuery
SQL DB	Amazon RDS	SQL Database	Cloud SQL
ETL	AWS Glue	Data Factory	Cloud Data Fusion
ML	SageMaker	Machine Learning	Vertex AI

Most of the Fortune 500 companies use cloud-based services. This makes cloud skills essential for data science and related roles. Here is a brief sketch of how a data scientist named Carla might work in the cloud: Carla arrives at the office ready to test a new predictor in a recommender system that her team built. She logs into the AWS Management Console and launches SageMaker Studio. She opens a notebook and selects the desired EC2 compute instances. Since she will be training a large machine learning model, she selects an instance from the P3 class. The datasets are saved in an S3 bucket (object file storage), and she loads them into the notebook. She pulls the latest project code from `GitHub`, creates a new branch for her experiment, and starts to modify code.

She runs experiments and saves her finalized model (to S3) and her code (to GitHub). When she's finished, Carla shuts down the SageMaker instance.

5.9 Chapter Summary

Hardware comprises the physical components of the computer, while software is the instructions to operate computers and execute tasks. The CPU is the brain of the computer which processes information and temporarily stores it in RAM. A GPU is a processor made of specialized cores which are excellent at performing parallel computations. This helps make them extremely useful for machine learning. RAM, or memory, is much faster than hard disk drive storage, but it is much more expensive. As a result, computers typically have much more HDD storage than RAM. Solid-state drives store data in integrated circuits, making them smaller and faster than HDD. The motherboard connects all of the principal hardware components.

The most important software on the computer is the operating system. Operating systems organize files into a directory structure. This structure is different in Windows versus non-Windows operating systems, and pathing looks slightly different.

The file format is the structure of a data file. Some common data formats are CSV, DOCX, and TXT. There are many specialized formats such as MP4, WAV, and PNG for media and Parquet for columnar data. Saving data in appropriate formats can speed up workloads significantly.

The shell supports interfacing with a computer. It is a computer program that takes commands, interprets them for the OS, and outputs the results. Users interact with the shell through a GUI or a command-line interface. Use of the CLI is generally through the terminal. Users enter commands at the prompt in the terminal.

Version control is essential for tracking file changes. The most widely used system for version control is Git. GitHub is a popular internet hosting service for software development and version control, and it is based on Git. We reviewed details of how to set up Git and GitHub, and how to execute common tasks like forking and cloning a repo. A fork of a repo will make an independent copy in the user's personal GitHub account. A clone of a repo will create a linked copy which will continue to sync with the original repo.

We walked through installing Anaconda, which provides a set of powerful, free tools for computing and data science. One of these tools is JupyterLab, which is a feature-rich notebook application and editing environment. This will be the primary method for interacting with files in this book. IDEs provide features for efficient code development, debugging, and testing. Spyder is the IDE that installs with Anaconda.

Cloud computing is a paradigm that supports scalable, secure, durable, pay-as-you-go pricing for computing services over the internet. Most large companies use cloud services for their computing, storage, and application needs. This translates to high demand for data scientists with cloud skills.

In the next chapter, we will start to leverage the course repo and dive into data processing with Python. This will include cleaning, enriching, and analyzing structured and unstructured data. The variety of datasets and tasks will increase your capabilities and knowledge.

5.10   Appendix: Going Further with Git and GitHub

This section provides details for other common tasks with `Git` and `GitHub`.

5.10.1   Syncing with the Upstream Repository

Making a fork of a repo produces an independent copy. This means that it will not receive changes to the original repo unless the sync process is run. This section walks through the sync process. Figure 5.10 shows the current repo setup, where the forked copy is linked to the clone on a local machine. In Panel A, the white background illustrates a scenario where the original has the latest changes, while the gray background indicates that the fork and clone are lagging behind. Running the sync will trigger an update between the original repo and the forked copy.

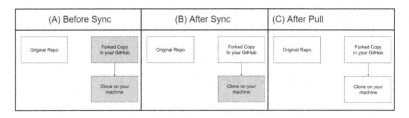

FIGURE 5.10: Repo Setup and Update Flow

Tip: **Avoiding a Merge Conflict.** If a change is made to a pre-existing file, it is a good idea to rename it before saving. If a change is made without renaming, a merge conflict will occur. This is because `Git` will see differences between the original and fork, and it won't know which version to keep. If this does happen, one solution is to move the changed file outside the repo and rerun the sync.

From the forked copy in your `GitHub` account, find the `Sync fork` button as shown in Figure 5.11. Clicking the button `Update branch` will trigger the update and merge changes.

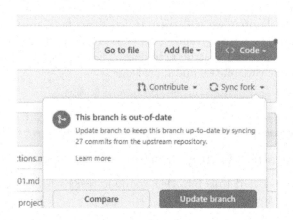

FIGURE 5.11: Sync the Fork

Following the sync, the forked copy will be current (Panel B).

Lastly, pull updates to the clone by returning to the terminal, changing directory to the repo, and running the command:

```
> git pull
```

This will pull the changes from the forked copy to your computer. This completes the sync (Panel C).

5.10.2   Initializing a Git Repo

For a new project to be version controlled, it must be initialized as a `Git` repo. The screenshot in Figure 5.12 illustrates the steps.

The `mkdir` command creates a new directory (folder) for the project, called `new_git_project`.

The `cd` command changes the directory to the new folder.

Finally, the `git init` command initializes the project as a `Git` repo (Figure 5.12).

FIGURE 5.12: Initialize a Repo

5.10.3 Tracking Changes

In this section, we track changes in a new file added to the repo named test_file.txt. To check the tracking status, the command `git status` can be typed at the prompt as in Figure 5.13.

Git mentions that to track this file, the command `git add` can be run, and so we do this.

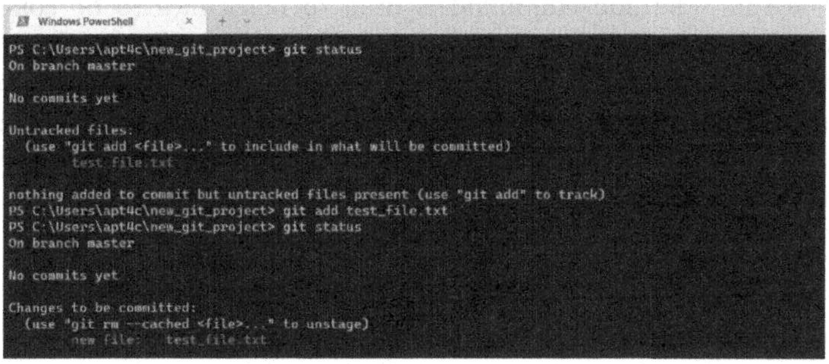

FIGURE 5.13: Check the Status

The file has "gone green," meaning that Git is now tracking it. Git manages tracking of files in a staging area. One or more files can be added to this staging area. For example, to add all files with extension txt, a file name pattern with the wildcard operator * can be used like this:

```
> git add *.txt
```

File changes in the staging area can be finalized with the commit command as shown in Figure 5.14.

The command includes a flag which appends a message to the commit, where the flag is -m and the message is

```
'added test_file'
```

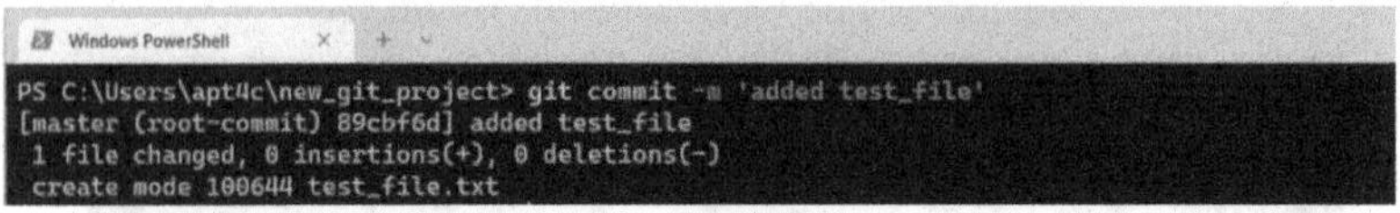

FIGURE 5.14: Running a Commit

A useful message will describe what the commit did. Lastly, running `git status` again will show that the `Git` staging area is empty and the commit is complete, as in Figure 5.15.

FIGURE 5.15: Clean Working Directory

New users to `Git` frequently ask why separate `add` and `commit` commands need to be run. The `add` command is used to mark files to be committed, which is called *staging*. It may happen that files are not all staged at once, or that some files are removed from staging. Once the user is ready, the `commit` command will save and track the changes. Using the analogy of a flight, the `add` command gets people onto the plane, and the `commit` command is akin to takeoff.

5.11 Exercises

Exercises with solutions are marked with Ⓢ. Solutions can be found in the `book_materials` folder of the course repo at this location:

https://github.com/PredictioNN/intro_data_science_course/

1. Ⓢ Explain the difference between hardware and software.

2. Explain the difference between a CPU and a GPU.

3. Ⓢ Why is data retrieval on a hard disk drive slower than retrieval in RAM?

4. Ⓢ What is a computer core?

5. Read about pipe-delimited files. Why are these sometimes used instead of CSV files?

6. ⓈExplain one or more benefits of using the command-line interface instead of a GUI.

7. ⓈWhat is the most important software on a computer?

8. Explain the difference between the Linux commands ls and ls -la.

9. Explain version control and why it is important.

10. ⓈTrue or False: You are working on a code project by yourself. To track file changes, you will need to use GitHub.

11. You are working with a repo on GitHub. What is the difference between cloning and forking a repo?

12. ⓈYou open a terminal and change the directory. Next, you enter git clone at the command line. What needs to be included to successfully clone a repo?

13. ⓈYou have created a GitHub account and you forked the course repo. What should you expect to happen next?

    a) The repo will be saved locally on your hard drive
    b) The repo will be copied to your GitHub account
    c) You will face a dreaded merge conflict with other users
    d) You will need to update your password

14. ⓈWhich of these are benefits of using Anaconda for Python coding? Select all that apply.

    a) It directly provides unlimited cloud computing with GPUs
    b) It includes many commonly used modules for computing
    c) It supports different operating systems
    d) It is free

15. Jupyter notebook files have which extension?

    a) .ipynb
    b) .PDF
    c) .JSON
    d) .CSV

6

Data Processing

After data is ingested and integrated, it needs to be processed from its raw form to make it more useful. There are many ways that processing might be done. Some tasks are done as part of a research and development function, such as interactive processing as part of a data science pipeline workflow. For example, a data scientist might load data from a file or database and write and run code to process the data for machine learning. Since the work is experimental and not required in real time, the speed of processing might not be critical. This kind of work is considered part of the *dev environment*.

On the other hand, some processing is time sensitive, as the data might feed a model that provides recommendations to customers in real time. If the customer needs to wait more than a few seconds for the response, she might abandon the application. This workload is running in the *production environment*, or *prod*. Data processing in prod needs to be automated and optimized for speed. It is unsafe for workers to test features in prod, as they might interrupt the product delivered to customers.

The methods and tools for working with data will depend on how the data is structured, which environment is in use, and other factors. In this chapter, we will build coding skills to process data in unstructured and structured formats. The examples will begin with smaller datasets and code samples, which will focus attention on learning the concepts and techniques.

The first dataset in our work will be an unstructured document in TXT format. The second dataset is structured and saved in CSV format. The third dataset is a larger CSV file in need of cleaning. It contains missing values and extreme observations. Since real-world datasets can be large, the cleaning and exploration should be done programmatically. Statistical and graphical summaries can be used to identify issues.

6.1 California Wildfires

Our first challenge is to load a news article about California wildfires, process it, and answer questions about it. Unstructured text is very common, so it will be valuable to have the skills to analyze this kind of data. In practice, there may be tens of thousands of such files that need to be processed and analyzed in a batch.

Datasets will be found in the course repo `intro_data_science_course` in one of two folders:

```
semester1/datasets
semester2/datasets
```

For this exercise, the file is saved in `semester1/datasets`.

It consists of a Wikipedia entry on California wildfires that was sourced from the URL

```
https://en.wikipedia.org/wiki/2021_California_wildfires
```

The notebook file we will work through is saved here:

```
semester1/week_03_04_data_types
```

and it is called `california_wildfire_text_processing.ipynb`.

The full path includes the path above with the file name like this:

```
semester1/week_03_04_data_types/
california_wildfire_text_processing.ipynb
```

Note that this single line of code extends to the next line due to margin constraints.

6.1.1 Running Python with the CLI

Note that while the code resides in a Jupyter notebook, this is not the only way to run the code. Python can also be run in a command-line interface. One important point is that the system needs to know the path to find Python. Since Anaconda applications have awareness of the path, they provide a straightforward means for accessing the Python CLI. We briefly review two access methods next.

One way to access Python at the command line is from the Anaconda Prompt. Windows users can search for the Anaconda Prompt by pressing Start and typing "Anaconda Prompt," while Mac users can search with Spotlight. Once the CLI launches, typing `python` should launch Python and provide the command-line prompt `>>>` for entering commands.

A second way to work with the Python CLI is through Jupyter Lab. From the Jupyter Lab interface, navigate to the top left and select File > New > Terminal. Once the CLI launches, typing `python` should launch Python and provide the command-line prompt.

6.1.2 Setting the Relative Path

We will need to tell the computer where to find the data. Earlier, we learned that it is best to specify a relative path from the working directory. To check the working directory, the following code can be run in a cell:

```
import os
os.getcwd()
```

This imports the `os` module and uses the `getcwd()` function to get the current working directory. We will talk more about Python functions later, but for now a quick definition will suffice:

A Python function takes zero or more inputs, runs a set of instructions, and returns zero or more outputs. At their best, they do useful work, and they are easily reusable.

Here is the output:

```
OUT:
'C:\\Users\\apt4c\\Documents\\PredictioNN_LLC\\repos
\\intro_data_science_course\\semester1\\week_03_04_data_types'
```

Since the notebook is saved in `week_03_04_data_types`, this working directory makes sense. Paths will be enclosed in quotes, and either single- or double-quotes will work.

Tip: The double backslashes appear because my Windows machine uses a single backslash for pathing, but Python uses that character as an escape character. To force the meaning of backslash, it adds the second backslash to each use.

To find the data file, we move like this from `week_03_04_data_types`

1. back up one directory
2. search in the `datasets` folder for the file `california_wildfire.txt`

The command `..` will back up one directory. Folders and files need to be separated with forward slash, and so we assemble the relative path like this:

```
PATH_TO_DATA = '../datasets/california_wildfire.txt'
```

Before loading in the data, there are two more concepts we need to cover: *variables* and *strings*.

6.1.3   Variables

We assigned the path to a variable named `PATH_TO_DATA`

Variables are used in computer science to store information to be used by the program. They are fundamentally different from variables in mathematics, and this can cause confusion for the uninitiated. In mathematics, variables are unknown quantities. In computer science, variables are known, and their values are stored in a memory address. If we want to see the memory address, we can use the `id()` function like this:

```
id(PATH_TO_DATA)
```

```
OUT:
2340478058544
```

The output is a memory address, which is not particularly illuminating on its own. However, sometimes this function is useful to check if two variables are truly different. If they both point to the same memory address, then they are the same.

The names of Python variables need to follow the rules:

• Names must start with a letter or underscore character

• Names must contain only letters, numbers, and the underscore character

• Reserved keywords such as **return** cannot be used

Also of note: names are case-sensitive. `VAR`, `VaR`, and `var` are different variables. Let's try to break a rule and see what happens:

```
4badvar = 5
```

```
OUT:
 Input In [11]
    4badvar = 5
       ^
SyntaxError: invalid syntax
```

Python throws a syntax error, due to an invalid variable name. In this case, it doesn't like that the variable begins with a number.

6.1.4   Strings

When we assigned the path variable, we issued:

```
PATH_TO_DATA = '../datasets/california_wildfire.txt'
```

The value of the variable is '../datasets/california_wildfire.txt'

This data type, enclosed in quotes, is a *string*. The string is one of the four primitive data types that we encountered earlier. It is a sequence of characters, and the data type commonly represents text. Python will allow single quotes, double quotes, or triple quotes around a string.

Once a string is created, it cannot be modified. If a different string is necessary, then it must be assigned to a new variable. A variable that cannot be changed is said to be *immutable*. This property is important, and it will arise later when we discuss other data types.

6.1.5   Importing Data

The code below sets the full path to the data file, and it uses functions to open the file and read from it. First, a quick observation: there are lines containing this character: #

Text following # is a comment to be ignored by the computer. Thoughtful comments are invaluable to humans, however, as they communicate the intention of code.

```
# set path to datafile
PATH_TO_DATA = '../datasets/california_wildfire.txt'

# open file and read into variable
f1 = open(PATH_TO_DATA,'r')
data = f1.read()
```

Now, if things went as intended, `data` contains the text in the file. Let's call the `print()` function on the data to print its contents. For brevity, we show the first few lines of output.

```
print(data)
```

OUT:
The 2021 California wildfire season was a series of wildfires
that burned across the U.S. state of California. By the end of
2021 a total of 8,835 fires were recorded, burning 2,568,948
acres (1,039,616 ha) across the state. Approximately 3,629
structures were damaged or destroyed by the wildfires, and at
least seven firefighters and two civilians were injured.

The wildfire season in California experienced an unusually
early start amid an ongoing drought and historically low
rainfall and reservoir levels.

This is exciting, as we were able to use Python to read data from a file into
a variable and print it to the console. Let's dig a little deeper on the `open()`
and `read()` functions. Different file formats can be opened and processed with
different functions. For this file in TXT format, the built-in `open()` function
takes a file path and a mode, and it returns a file object. This code uses the
read mode 'r'. To open a pre-existing file and write to the end of it, one could
use the append mode 'a' . Next, `read()` returns all file contents as a string.
For more details about `open()` and other built-in functions, please see the
Python documentation:

`https://docs.python.org/3/library/functions.html`

It is always a good idea to verify the data type of objects. Oftentimes when
errors arise, it is because our expectations are not aligned with reality: the
data is a different type, the shape of a matrix is different than we thought,
etc.

```
type(data)
```

```
OUT:
str
```

We see that data is a string, which is expected.

6.1.6   Text Processing

Next, let's process the article to make it more useful. We can split the data
on periods using the `split()` function. This will approximately segment the
article into sentences. We will save the result into a variable named `sent`, and
we will print it.

```
sent = data.split('.')
print(sent)
```

Here are the first few lines of output:

```
OUT:
['The 2021 California wildfire season was a series of wildfires
that burned across the U', 'S', ' state of California', ' By
the end of 2021 a total of 8,835 fires were recorded, burning
2,568,948 acres (1,039,616 ha) across the state', '
Approximately 3,629 structures were damaged or destroyed by the
wildfires, and at least seven firefighters and two civilians
```

were injured', \n\nThe wildfire season in California
experienced an unusually early start amid an ongoing drought
and historically low rainfall and reservoir levels',

Splitting on the periods removes them, and we see a few things that are new:

First, the fragments of text are wrapped in quotes, which indicates they are strings.

Second, there is an \n\n string. The '\n' character indicates a new line. Looking back at the original text, there is a gap between lines 6 and 8. The first '\n' ended line 6, and the second '\n' skipped line 7.

Third, the output begins with [. If we looked at all of the split text, we would see that it ends with]. Let's check the data type of sent to understand what is happening.

```
type(sent)
```

OUT:
list

It turns out that sent is a list of strings. We have just discovered a new data type. Unlike strings, the contents of a list can be changed. It is also possible to subset, or index into, a list to extract elements from it.

Let's select the first string, or element, from sent.

```
sent[0]
```

OUT:
'The 2021 California wildfire season was a series of wildfires
that burned across the U'

This gives back the first string.

Notice that the first string was stored in position zero. Python uses this convention for all data types storing a collection of elements. sent[1] will contain the second element, and if there are 10 elements, then the last one will be stored in sent[9].

Additionally, [] is called the *subset operator*. We will use it to subset other data types as well.

Next, let's do something quantitative. How many strings, or sentence fragments, are in the list? We can use the len() function to count the number of elements in the list.

```
len(sent)
```

```
OUT:
21
```

Running `type(21)` will return `int`, which indicates that the value is an integer.

Breaking a document down into its words, or *word tokenization*, is a common and useful step in language processing. This can be useful for inferring the topic, for instance. Next, we will tokenize the first string in `sent`. We can use the `split()` function to split on spaces.

```
tokens = sent[0].split(' ')
print(tokens)
```

```
OUT:
['The','2021','California','wildfire','season','was','a',
'series','of','wildfires','that','burned','across','the','U']
```

Notice what this did: it produced a list of strings, where each string is a word. Python datatypes can be recognized by their punctuation; strings are enclosed in quotes, and lists are enclosed in square brackets.

Next, we would like to associate each word with its position in the sentence. This is useful if we want to incorporate text data in a model, for example. We can't directly feed text into a model, so we need to represent it numerically.

Here is what we would like to build:

```
word_index = {'The': 0, '2021': 1, 'California': 2, 'wildfire':
3, 'season': 4, 'was': 5, 'a': 6, 'series': 7, 'of': 8,
'wildfires': 9, 'that': 10, 'burned': 11, 'across': 12, 'the':
13, 'U': 14}
```

This Python object is called a *dictionary* and it stores key:value pairs. The keys in this case are strings, and the values are integers indicating positions, starting from zero. The code snippet below provides a solution.

```
# create empty dictionary
word_index = {}

for ix, token in enumerate(tokens):
        word_index[token] = ix

print(word_index)
```

```
OUT:
```

```
{'The': 0, '2021': 1, 'California': 2, 'wildfire': 3, 'season':
4, 'was': 5, 'a': 6, 'series': 7, 'of': 8, 'wildfires': 9,
'that': 10, 'burned': 11, 'across': 12, 'the': 13, 'U': 14}
```

Dictionaries use curly brackets to wrap their data, and the first command creates an empty dictionary.

Next, a *for loop* passes over each word in the list. The `enumerate()` function will produce two things for each step of the loop:

The `ix` variable acts as a counter. For the first step, it will hold value 0. On the next step, it will be 1, then 2, and so on. The name `ix` is completely arbitrary.

The `token` variable will hold the data at each step. For the first step, it will hold 'The', on the second step it will hold '2021', and so on. The name `token` is also arbitrary.

Let's take the first step in the for loop. At this step, `ix` = 0 and `token` = 'The'

The next thing that happens is that we store the value into the dictionary at a precise location.

This compact little statement:

```
word_index[token] = ix
```

gives the instruction to store the value 0 into the dictionary where the key is "The".

If we wanted to check the value in the dictionary when the key is "The" we could run:

```
word_index['The']
```

and the output would be 0.

The loop then does the identical work for each of the other words, storing the words as the keys and the counters as the values.

After the for loop processes each word in the list, it stops. Before continuing, please convince yourself that this works, and run the code. One thing that often helps is to print intermediate results. Here is a snippet for testing, which will print the counter and value on each iteration:

```
for ix, token in enumerate(tokens):
        print(ix, token)
        word_index[token] = ix
```

The last task in this exercise is to extract a string with duplicate words, and run a command to determine the unique words. A Python data type called a *set* will do this for us. A set holds a unique collection of objects. If the objects are not unique, it deduplicates them to make them unique. It will be highly useful in data science to work with unique objects, such as unique users and unique products.

Now, let's subset the list and produce word tokens:

```
sent[9].split(' ')
```

```
OUT:
['\n\nThe','long','term','trend','is','that','wildfires',
'in','the', 'state','are','increasing','due','to','climate',
'change','in','California']
```

Notice the list isn't unique as 'in' appears twice.

The solution is quick: Wrap `set()` around the statement like this:

```
set(sent[9].split(' '))
```

```
OUT:
{'\n\nThe','California','are','change','climate',
'due','in','increasing','is','long','state','term',
'that','the','to','trend','wildfires'}
```

This object is a set, and the words are unique. Notice that the order is shuffled; sets cannot be used to maintain order.

6.1.7 Getting Help

This section introduced several built-in functions, but there are many others. To get quick help on a Python object – including functions, modules, and methods – run the `help()` command:

```
help(len)
```

```
OUT:
len(obj, /)
    Return the number of items in a container.
```

In addition, Python has excellent options for help on the web, including its own documentation (https://docs.python.org/) and Stack Overflow. The truth is that data scientists and software developers do a lot of web searching when writing code. They get more efficient in their searching as they understand the fundamentals and relevant keywords.

6.2 Counting Leopards

In this exercise, we will use Python to analyze structured data on the Amur leopard, which is the rarest cat in the world. Currently, there are fewer than 100 Amur leopards left in the wild, and they can be found in China and Russia. They have beautiful, spotted fur and can leap 19 feet horizontally and up to 10 feet vertically.

Counting animals, or things in general, can be very challenging. Data scientists need to ensure that data is accurate, because no analysis can overcome bad data. In the case of the Amur leopard, it can cross borders between countries, and this can result in overstated population counts and distrust. A camera trap is a camera that is automatically triggered by a change in activity, such as the presence of an animal or human. Accurately counting the leopards using camera traps with a coordinated effort across country boundaries is the subject of [23]. The paper is also the origin of this data source.

Let's set the path to the data and load the CSV file. To work with the structured (or *tabular*) data, we will import the **pandas** module. Some modules are used so often that they are given an alias, or nickname. Using the alias **pd** for **pandas**, we can reference its functions.

In this case, **pd.read_csv()** allows for reading the data into a dataframe object. Commonly referenced as **df**, the dataframe provides a wealth of functionality for data wrangling and analysis. The **pandas** module is very powerful, and we will work through some of its functionality. For a more thorough reference, see [24]. Note that in some tables, the **index** column is omitted due to margin constraints.

```
import pandas as pd

# set path to datafile
PATH_TO_DATA = '../datasets/amur_leopards.csv'

# read in the CSV file and print the data
df = pd.read_csv(PATH_TO_DATA)
df
```

year	location_captured	females	males	cubs	unknown_sex	total
2014	China	14	11	2	0	27
2014	Russia	25	21	3	2	51
2014	China & Russia	33	24	5	2	64
2015	China	10	12	0	0	22
2015	Russia	24	20	8	3	55
2015	China & Russia	31	25	8	3	67

The data represents the counts and sex of Amur leopards captured by camera trap surveys in China and Russia. In some cases, the China and Russia totals exceed China & Russia, and this is because some leopards were observed in both countries. The importance of understanding what the data means cannot be overstated.

Next, we're going to run through a series of tasks with the data to build our skillset.

6.2.1 Extracting DataFrame Attributes

We can extract the data like this:

```
df.values
```

```
OUT:
array([[2014, 'China', 14, 11, 2, 0, 27],
       [2014, 'Russia', 25, 21, 3, 2, 51],
       [2014, 'China & Russia', 33, 24, 5, 2, 64],
       [2015, 'China', 10, 12, 0, 0, 22],
       [2015, 'Russia', 24, 20, 8, 3, 55],
       [2015, 'China & Russia', 31, 25, 8, 3, 67]],
      dtype=object)
```

Checking the data type with `type(df.values)` yields a `numpy.ndarray`.

We can list the columns with:

```
df.columns
```

```
OUT:
Index(['year', 'location_captured', 'females', 'males', 'cubs',
'unknown_sex','total'],dtype='object')
```

If we check the type of this object, it is `pandas.core.indexes.base.Index`, which is a specialized `pandas` object. Sometimes we will want this as a list. We can convert or cast the index to a list like this:

```
list(df.columns)
```

OUT:
['year','location_captured','females','males','cubs',
'unknown_sex','total']

In the first column of the dataframe is a set of values called indexes. These are generated by **pandas**. The index column can be used to reference the rows, and they can be extracted by running:

```
df.index
```

OUT:
RangeIndex(start=0, stop=6, step=1)

We can subset on the first row like this:

```
df.loc[0]
```

OUT:

```
year                2014
location_captured   China
females             14
males               11
cubs                2
unknown_sex         0
total               27
Name: 0, dtype: object
```

When one row or column is returned, its data type is actually not a **pandas** dataframe, but rather a **pandas** Series.

6.2.2 Subsetting

Subsetting on multiple rows will produce a dataframe. Here, we select rows 0 through 2:

```
df.loc[0:2]
```

OUT:

year	location_captured	females	males	cubs	unknown_sex	total
2014	China	14	11	2	0	27
2014	Russia	25	21	3	2	51
2014	China & Russia	33	24	5	2	64

Select the year column:

```
df['year']
```

```
OUT:
0    2014
1    2014
2    2014
3    2015
4    2015
5    2015
Name: year, dtype: int64
```

Select multiple columns using a list of strings:

```
df[['year','location_captured']]
```

OUT:

index	year	location_captured
0	2014	China
1	2014	Russia
2	2014	China & Russia
3	2015	China
4	2015	Russia
5	2015	China & Russia

Next, let's get a little fancier. We will apply some filtering to the data.

Select all rows where `location_captured` is China:

```
df[df.location_captured == 'China']
```

OUT:

year	location_captured	females	males	cubs	unknown_sex	total
2014	China	14	11	2	0	27
2015	China	10	12	0	0	22

Let's break this down. The expression `df.location_captured == 'China'` verifies if the `location_captured` value is equal to 'China' for each row in the dataframe. The result is a series of Boolean values, or True/False values. These data types, which are `bool`, are new to us.

```
0    True
1    False
2    False
3    True
4    False
5    False
```

These Booleans are used to filter the rows of the dataframe df. This means that only rows where the value is True will be returned, which explains why rows 0 and 3 were output.

How can we determine which locations captured less than 15 females in 2015? We can filter on multiple conditions to answer this question, since we are applying criteria on females and year.

```
df[(df.females < 15) & (df.year == 2015)]
```

OUT:

year	location_captured	females	males	cubs	unknown_sex	total
2015	China	10	12	0	0	22

Notice that & is the AND operator (note that | is the OR operator).

This gets the correct row, but it returns unnecessary columns. Let's adjust the statement to select only the location_captured column.

```
df[(df.females < 15) & (df.year == 2015)]['location_captured']
OUT:
3    China
```

This gets the desired answer. The syntax we used to subset on rows and columns was:

```
dataframe[row_condition][column_condition]
```

The column_condition was simple: it was the column name. If we want several columns, we can put them in a list.

We have been subsetting data by using dataframe methods. For more complex subsetting, it often makes sense to use SQL, which is shorthand for Structured Query Language. We will dive into SQL later in the book.

6.2.3 Creating and Appending New Columns

Next, we would like to do a simple calculation and store the results in the dataframe: for each row, sum the number of females and males (these are the adults with known sex). Since the operation is simple, we can write an expression involving the necessary columns, and assign the result to a new variable with the equals operator like this:

```
df['f_and_m'] = df['females'] + df['males']
```

Alternatively, we can use a dot notation on the right, which is shorter and often preferred:

```
df['f_and_m'] = df.females + df.males
```

OUT:

year	location_captured	fems*	males	cubs	unk_sex*	total	f_and_m
2014	China	14	11	2	0	27	25
2014	Russia	25	21	3	2	51	46
2014	China & Russia	33	24	5	2	64	57
2015	China	10	12	0	0	22	22
2015	Russia	24	20	8	3	55	44
2015	China & Russia	31	25	8	3	67	56

* `females` and `unknown_sex` are abbreviated to fit. `index` is suppressed.

`Pandas` performed this operation on each row, or row-wise, which was very convenient. For more complex row-wise operations, we can define a function and use the `apply()` function. We will discuss this later.

We cannot use the dot notation when creating a new variable; this is not valid syntax. Watch what happens:

```
df.f_and_m = df.females + df.males
```

OUT:
```
UserWarning: Pandas doesn't allow columns to be created via a
new attribute name - see https://pandas.pydata.org/pandas-
docs/stable/indexing.html#attribute-access
df.f_and_m = df.females + df.males
```

We will do two final things with this dataset: sort the data by the new column, and save it to disk.

6.2.4   Sorting

We can sort by one or more columns by passing a list of the column names to the `sort_values()`. Sorting will be temporary unless we assign the result to a variable or include the parameter `inplace=True`. Let's sort by our new column from high to low, or descending. To sort descending, set the parameter `ascending=False`.

```
df.sort_values(['f_and_m'], ascending=False,
inplace=True)
```

OUT:

year	location_captured	fems	males	cubs	unk_sex	total	f_and_m
2014	China & Russia	33	24	5	2	64	57
2015	China & Russia	31	25	8	3	67	56
2014	Russia	25	21	3	2	51	46
2015	Russia	24	20	8	3	55	44
2014	China	14	11	2	0	27	25
2015	China	10	12	0	0	22	22

6.2.5 Saving the DataFrame

Now that we have our final table, we can save it to a file to avoid repeating our work later. The code below creates a variable with the desired output path. As a reminder, the '`./`' prefix says: save the file in the current directory. The to_csv() function will save the dataframe to a CSV file.

```
PATH_TO_OUTFILE = './amur_leopards_final.csv'
df.to_csv(PATH_TO_OUTFILE)
```

6.3 Patient Blood Pressure

6.3.1 Data Validation

It is always important to explore and validate data. There might be errors lurking, and it is essential to resolve them before doing important analysis. In this exercise, we will analyze some fictitious patient blood pressure data. It consists of repeated measurements of systolic and diastolic pressure on multiple patients. As we will discover, there are some thorny issues to tackle. We will use **pandas** again, as the file is in CSV format. Let's read in the file and look at the first few records.

```
import pandas as pd

# set path to datafile
PATH_TO_DATA = '../datasets/blood_pressure.csv'

# read in the CSV file and print the data
df = pd.read_csv(PATH_TO_DATA)
df.head()
```

OUT:

index	patientid	date	bp_systolic	bp_diastolic
0	1	02/01/22	120	76.0
1	1	02/02/22	127	75.0
2	1	02/03/22	127	70.0
3	1	02/04/22	127	76.0
4	1	02/05/22	−20	74.0

Notice there are some irregularities with data types. The systolic blood pressure column seems to contain integers, while the diastolic blood pressure column looks like floating point values, or *floats*. For consistency and to avoid issues later, we should investigate the data further, think about which type makes sense, and make the appropriate conversions.

We can check the data types of the dataframe by retrieving its `dtypes` attribute like this:

```
df.dtypes
```

```
OUT:
patientid          int64
date               object
bp_systolic        int64
bp_diastolic       float64
```

This confirms our suspicion. One thing to note is that a missing value in a column will result in the entire column having the float data type. Let's summarize the data next by calling `describe()` which is a fast way to compute column counts, percentiles, means, and standard deviations (std).

```
df.describe()
```

OUT:

	patientid	bp_systolic	bp_diastolic
count	99.000000	99.000000	98.000000
mean	2.595960	124.171717	75.653061
std	0.978537	16.744198	5.143512
min	1.000000	−20.000000	70.000000
25%	2.000000	122.000000	73.250000
50%	3.000000	125.000000	76.000000
75%	3.000000	128.000000	77.000000
max	4.000000	200.000000	120.000000

Notice there are 99 systolic blood pressure values, but only 98 diastolic values. Let's check for missing values by calling the `isnull()` function on the bp_diastolic column:

```
df[df.bp_diastolic.isnull() == True]
```

This places a condition on the rows: for any row where `bp_diastolic` is missing, it will result in True. The dataframe is then filtered by the resulting Booleans, and all rows with the True condition are returned.

OUT:

index	patientid	date	bp_systolic	bp_diastolic
6	1	02/07/22	121	NaN

This isolates the row with the missing value (coded as NaN) and explains why the column contains floats. Next, we should decide how to *impute*, or fill in, this value.

6.3.2 Imputation

`Pandas` offers several different methods for easily imputing values, such as replacement with the median, the mean, a value of choice, or the last non-missing value. Like all things with data science, it is important to consider what makes sense conceptually. For any variable that is imputed, the analyst needs to ask what is appropriate. It may be the case that each variable needs its own imputation method.

For this dataset, there are several patients, and each patient has several observations.[1] It might make more sense to use each patient's data for imputation, rather than computing a statistic across all patients. The choice made here is to impute with the last non-missing value, or *last value carried forward* (LVCF). This might not be the true measurement at that time point, but it may be reasonable. The `pandas ffill()` function will accomplish this, and passing the parameter `inplace=True` will update the value in the dataframe.

```
df.ffill(inplace=True)
```

Checking for missing again returns no records, meaning that imputation was successful.

```
df[df.bp_diastolic.isnull() == True]
```

OUT:

index	patientid	date	bp_systolic	bp_diastolic

Looking at records around the previously missing row confirms the LVCF:

---

[1]The number of unique patients can be computed with `len(df.patientid.unique())`

```
df.loc[5:7]
```

OUT:

index	patientid	date	bp_systolic	bp_diastolic
5	1	02/06/22	129	75.0
6	1	02/07/22	121	75.0
7	1	02/08/22	125	74.0

6.3.3   Data Type Conversion

Next, let's make the blood pressure data types match. In this case, since we don't measure blood pressure to decimal precision, both systolic and diastolic can be integer types. This will mean converting bp_diastolic to integer, which can be done like this:

```
df = df.astype({'bp_diastolic':np.int32})
```

We used the astype() function, passing a dictionary with the column name as key and the required data type as value. We could have passed additional key:value pairs to convert multiple columns if we wished.

Now we check the data types, and we see that bp_diastolic has the desired integer type.

```
df.dtypes
```

```
OUT:
patientid        int64
date             object
bp_systolic      int64
bp_diastolic     int32
```

The data type conversion produces a new dataframe. We can see this by calling id() before and after the assignment and noticing that the dataframe memory addresses differ.

```
id(df)
2641496136432
```

```
df = df.astype({'bp_diastolic':np.int32})
```

```
id(df)
2641537679264
```

6.3.4 Extreme Observations

From the statistical summary produced by `describe()`, `bp_systolic` holds a negative value and a very high value of 200. Negative blood pressures do not make clinical sense, while the value of 200 is possible but concerning. There are different options for handling the negative value. Here, we set a rule where we temporarily replace negative values with missing, and then fill missing values with the previous measurement (that is, the LVCF).

```
df.loc[df.bp_systolic < 0, 'bp_systolic'] = np.nan
df.bp_systolic.ffill(inplace=True)
```

Next, let's turn to the patient with the systolic measurement of 200. Unlike the prior case, we cannot simply reject it as incorrect. It is best to check for contextual information before taking action. If this data was from an electronic health record system, we might check for notes. In this case, the correspondingly high diastolic measurement of 120 offers supporting evidence. Acting conservatively, we leave the outlier measurement in place.

6.4 Chapter Summary

Variables store information to be used by a program, and their values are stored in a memory address. Strings are immutable data types that hold a sequence of characters. Sets, lists, and dictionaries are data types that store collections of data. Sets store unique objects. Dictionaries use key:value pairs to store their data.

We processed unstructured data, which involved the use of string operations. For the structured data that we processed, we used **pandas** dataframes. The dataframe provides nice functionality for manipulating rows and columns of data. In particular, we learned many techniques to perform common operations including subsetting, filtering, imputing, and handling extreme observations. For going deeper, these notebooks from the course repo are recommended:

```
semester1/week_05_08_data_clean_prepare/pandas_dataframes.ipynb
semester1/week_05_08_data_clean_prepare/pandas_dataframes2.ipynb
```

Another helpful resource is the "10 minutes to pandas" guide, which can be found here:

```
https://pandas.pydata.org/docs/user_guide/10min.html
```

We have made a lot of progress in building data processing skills. In the next chapter, we will learn about storing and retrieving data. The data might have been processed and saved by a data scientist, or it might have been done automatically as part of a system. Databases are essential to data storage, and we will learn what they do and how to work with them.

6.5 Exercises

Exercises with solutions are marked with (S). Solutions can be found in the **book_materials** folder of the course repo at this location:

https://github.com/PredictioNN/intro_data_science_course/

1. (S) Explain the difference between an absolute path and a relative path. Why is it preferable to use an absolute path?

2. You are working with a file that has this path:

   'C:\Users\abc\Documents\some_file.csv'

   Is this computer running a Windows operating system? Explain your answer.

3. (S) In our text analytics, we split the article on periods to extract sentences. Can you think of any issues with segmenting sentences by splitting on periods? Explain your answer.

4. For each of the type conversions listed, indicate if they will succeed or fail:

   a) Converting from a string to an integer
   b) Converting from a string to a float
   c) Converting from an integer to a string
   d) Converting from an integer to a float

5. (S) For the wildfire article, tokenize it on spaces and calculate the average length of the tokens. For example, an article consisting of "california wildfire" would produce the tokens "california" (length=10) and "wildfire" (length=8), for an average token length of 9.

6. Provide an example of data that should be stored in a dictionary instead of a list.

7. (S) True or False: The **pandas** dataframe is most useful for unstructured text data.

8. (S) The wildfire exercise included the creation of a dictionary named **word_index** that held strings as keys and positions as values. Write

a function that takes `word_index` as input and returns a new dictionary where the keys and values are reversed. That is, the keys should hold the positions and the values should hold the strings.

9. Describe two ways that rows can be selected from a dataframe.

10. For the leopard dataset, there was a point made about understanding what the data represents. Explain this point.

11. Ⓢ Suppose that you've extracted the values from a `pandas` dataframe `df` with the `df.values` attribute. What is the data type of these values?

12. For the blood pressure dataset, write code to count the number of records for each patient. Avoid using a loop.

13. Ⓢ For the `patientid` variable in the blood pressure dataset, are the percentile statistics useful? Explain your answer.

14. Ⓢ A data scientist uses the `ffill()` function to impute some missing data in a dataframe column. Provide an example where this practice would not work well and explain your reasoning.

15. A data scientist notices patient blood pressure measurements which are negative. She reasons that such values cannot be negative and updates them to a value of zero. Is this a good practice? Explain your reasoning.

7

Data Storage and Retrieval

In this chapter, we will learn about databases and how to use them to store and retrieve data. We can think of a database as a tool that supports the creation (C), retrieval (R), updating (U), and deletion (D) of data. These operations are called the CRUD operations, and they provide minimal requirements for a database. For example, our brains qualify as a database under this definition. Unfortunately, brains tend to forget things that aren't sufficiently used. For a database to be trusted and useful, it needs to offer:

- Perfect retrieval, so that every piece of data can be accessed

- Speed, so that all operations are fast ... even at scale

- Consistency, so that if two users make the same request at the same time, they see the same result

- Ease of error correction

Different database systems offer different features. For massively data-intensive applications, it is not possible to perfectly satisfy all of the features, and tradeoffs must be made. It is useful, however, to keep CRUD in mind, and the considerations for a useful database.

We will not have the space here to provide a detailed treatment of databases, nor to discuss proper design. An excellent book for this purpose is [25]. Here, we will approach databases from the likely viewpoint of a data scientist new to an organization. In this scenario, a database (or collection of databases) is already in place. The structure may have been designed by a data architect and built by a data engineer. The data may have been saved by an automated process. The data scientist is then required to get access to the data and start retrieving it. For this to happen, the data scientist must learn a programming language to query the data.

There is a segmentation of database types into *relational databases* and *non-relational databases*. At the time of this writing, most databases are still relational databases. The language for querying relational databases is SQL. Non-relational databases are also called NoSQL databases, which indicates that SQL is not used for queries. The relational database requires that the data is structured according to a pre-defined schema. Much of the data in the modern world – including documents, images, and videos – is not amenable to schemas and such structuring. To accommodate the wide variety of data, specialized NoSQL databases have emerged, including:

- Key-value stores: a dictionary structure that allows for flexible, fast storage

- Graph databases: data is stored as nodes (which model objects) and edges (which model relationships)

- Document databases

The NoSQL databases are very interesting and useful, but since they are specialized, they are out of scope for this book. References for further exploration can be found at the end of this chapter.

For the remainder of this chapter, we will focus on studying relational databases. SQL will be introduced, and we will get hands-on experience working with a database through Python. This will provide an introduction and some practice. Databases can be a component of much larger storage systems, such as a *data warehouse* or a *data lakehouse*. We will briefly learn about these objects.

7.1   Relational Databases

One of the essential building blocks of a relational database is a table. A database table consists of records (also called rows or tuples, as a reminder) and fields (also called columns or attributes). A natural question is whether a single table should hold all of the data. Storing all of the data in one table can introduce risk, such as inconsistency. Consider the table below, which shows apps used by students on various devices. The field named status is used to categorize a user's total app use across devices.

Student App Use

name	device	app	status	sessions
Callie	phone	Sharpen	Moderate User	22
Callie	tablet	Sharpen	Moderate User	30
James	laptop	PebbleGo	Power User	40
Avery	phone	Duolingo	New User	2

For example, Callie had 30 sessions with the Sharpen app on her tablet, and 52 sessions across all devices. Based on her activity, she was categorized as a Moderate User. Suppose that over the next week, she has 15 more sessions on her tablet. Let us further suppose that a user with 60 or more sessions in Sharpen across all devices is categorized as a Power User. In this case, Callie's status should be updated to Power User. Imagine that the update is made to the second record of the table, but not the first record. This introduces inconsistent data, rendering the database less useful. At the heart of the problem is the duplication of data. If we maintain user status in a separate table with a single record for each user, this problem can be avoided. Here is what a database using two tables can look like:

Student App Use

name	device	app	sessions
Callie	phone	Sharpen	22
Callie	tablet	Sharpen	30
James	laptop	PebbleGo	40
Avery	phone	Duolingo	2

Student Status

name	status
Callie	Moderate User
James	Power User
Avery	New User

Given this new structure, there is no duplication of user status. This reduces the chance of data inconsistency. In fact, the first table can be broken down into smaller tables. We won't do this here, but the important point is that each concept should be stored in a separate table. The user's status is a concept, and this data should be stored in one table. Another concept is device, and the unique devices should be stored in another table. In this way, the structure is cleaner and less prone to error.

There is complexity that is introduced when breaking the data into one table per concept: when the required data is spread across multiple tables, the tables need to be joined. If the required data is in three tables A, B, and C, then we need a join between A and B, and a second join between B and C. This idea is fundamental to relational databases – it is what makes them "relational." When joins cannot be done, or they aren't needed, then a relational database is not the right tool for the job.

Figure 7.1 shows a database containing two tables. The `Customer` table stores all of the customer data, and the `Invoice` table stores all of the invoices. This follows the "one table per concept" principle of relational databases. If we want to combine the information, this can be done by joining the records on a common field. In this case, `ssn` appears in both tables, and it can be used for the join.

7.1.1 Primary Key

The field `ssn` is a *primary key* (PK) in the `Customer` table. A primary key:

- Must have unique values to tell the records apart

| **Customer** |
| ssn (primary key) |
| first_name |
| last_name |
| email |

| **Invoice** |
| order_number |
| price |
| quantity |
| ssn |

FIGURE 7.1: Database with Two Tables

- Can be used to join other tables

- Can consist of multiple fields, in which case it is called a *composite key*

Let's look at two examples. In Example 1, the first two records have matching first names and last names. Based on names, the records are not unique. However, the social security number can uniquely identify the records, and it can be used as the primary key.

Example 1: Social Security as Primary Key

ssn	first_name	last_name	email
123-45-6789	Sarah	Connor	sconnor@ymail.com
544-24-8764	Sarah	Connor	sjconnor@gmail.com
112-37-9080	Rob	Frey	bfrey@gmail.com

In Example 2, the table contains sample records of daily stock prices. Each record contains a stock ticker, date, and adjusted closing price. It is common to create tall tables like this for storing time series data. It will not be possible to use `ticker` as the primary key, but `ticker` and `date` together can uniquely define a record. We can form a composite key from these two fields. The PK abbreviation in the table next to `ticker` and `date` indicates that these fields form the composite key.

Example 2: Composite Key

ticker (PK)	date (PK)	adjusted_close
XYZ	2012-01-03	0.85
XYZ	2012-01-04	0.78
XYZ	2012-01-05	0.79

7.1.2 Foreign Key

A *foreign key* (FK) is a field in one table that references the primary key in another table. Foreign keys can help validate data. Consider a database containing two tables as shown in Figure 7.2. The Product table contains a field called product_name. This field is the primary key. A second table called Pricing contains the products offered and their current prices. The fields in this table are product_name (the foreign key) and price. Suppose that a user prepares pricing data to be inserted into Pricing, and one of the product names is incorrect. Since product_name is a foreign key in this table, the insertion will fail due to a violation of *referential integrity*. This is because the permitted values of product_name are limited to values in the Product table.

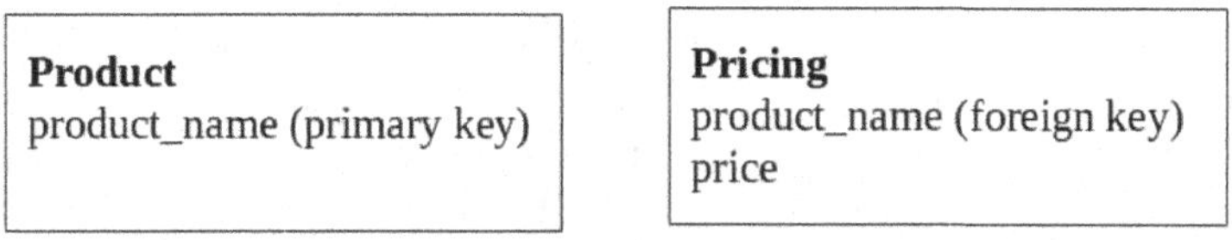

Product	**Pricing**
product_name (primary key)	product_name (foreign key) price

FIGURE 7.2: Primary and Foreign Key

This is enough relational database background to get us started. In the next section, we will work with the SQLite relational database management system (RDBMS). There are several other excellent RDBMSs available, such as MySQL, PostgreSQL, Oracle, and SQL Server. Some are freely available, such as PostgreSQL, while others are paid, such as Oracle and SQL Server.

7.2 SQL

SQL was developed in the 1970s at IBM, and it is the universally accepted way to work with relational databases. Users submit a *query* to accomplish tasks such as creating and deleting tables, inserting data, and retrieving data. Selecting all of the data from a table, for example, can be done like this:

```
SELECT * from my_table;
```

There is a small set of very common commands like SELECT which are written in uppercase by convention. The queries generally follow a style to make them more readable, such as:

```
SELECT
    product_id,
```

```
      product_name
FROM my_table
ORDER BY product_name;
```

This query retrieves data from the columns `product_id` and `product_name` across all rows of the table, sorting by `product_name`. In the sections that follow, we will review illustrations of SQL commands in action, along with detailed explanations. Note that different relational database systems may use different extensions, and the syntax may differ slightly. Additionally, when using an API in Python, the API will have its own conventions. We will see this in the next section.

Practical Matters

As a data scientist, it is extremely important to build skill in SQL, since data manipulation is a large part of the job. Many companies like to include technical interviews where they expect interviewees to write correct `SQL` code to munge data. The way to get good at SQL is to practice on challenging code examples, breaking them down until they seem obvious. For students feeling discouraged by a complex query, it can be helpful to take a step back and strategize on how to deconstruct it into subproblems. Then the subproblems can be pieced back together.

All `SQL` should be stored and version controlled, as it is code. For example, the commands to build the tables and insert the data might be saved in files on `GitHub`. This allows stakeholders to understand what was done, and to replicate the work if needed.

Enough background info, now let's get to work! In the next section, we will use `SQLite`. We will walk through creating a database and interacting with it using Python. Specifically, we will write and execute `SQL` in Python to do these things:

- Create a table

- Insert records into the table

- Query the data

7.3   Music Query: Single Table

In this section, we will learn and apply `SQL` on a single table of music artist data. We will begin by downloading and installing `SQLite`. Then we will launch and work with `SQLite` from a terminal. Lastly, we will work with `SQLite` from Python. In the next section, we will extend this to working with multiple tables. The course repo contains notebook demonstrations using different data for variety. The notebooks are located here:

`semester2/week_09_11_relational_databases_and_sql/`

I. Download and Installation

Download SQLite from `https://www.sqlite.org/download.html`

Windows users will install:

`sqlite-tools-win32-x86-[some_number].zip`

Mac users will install:

`sqlite-tools-osx-x86-[some_number].zip`

where [`some_number`] is a version number that changes over time.

If the `sqlite` folder name is long, it can be shortened to `sqlite`, for example. It is important to make note of where the database is saved. This location is the database path.

On a Windows machine, the path might look something like:

`C:\Users\apt4c\Documents\database\sqlite`

On a Mac, it might look something like:

`/user/apt4c/Documents/database/sqlite`

II. Launching Terminal and Working with SQLite

The next step after installation is to launch a terminal. Windows users will press the `START` key and type `CMD`.

From the terminal, it will be necessary to change the directory to the database path:

```
> cd C:\Users\apt4c\Documents\database\sqlite
```

where:

'>' is the prompt and cd is the change directory command.

Next, we create a database called musicians with this command:

```
> sqlite3 musicians.db
```

The output will look something like this:

```
SQLite version 3.39.4 2022-09-29 15:55:41
Enter ".help" for usage hints.
```

The prompt will change to this:

```
sqlite3>
```

Next, list the databases with this command:

```
sqlite> .databases
```

```
OUT:
main: C:\Users\apt4c\Documents\database\sqlite\musicians.db r/w
```

The output should include musicians.db. In the musicians folder, there should be a file called musicians.db which stores the database.

It is possible to interact with the database in terminal, but here we will switch to Python.

III. Working with SQLite from Python

Python has APIs for working with databases. Here we use the SQLite API. The API uses specific code that wraps around the SQL query for execution. To distinguish the SQL code from the wrapper, the query will be defined in each code snippet. We begin by importing the module and setting the relative path to the database. This assumes that the working directory is

```
C:\Users\apt4c\Documents\database
```

```
# import the SQLite API
import sqlite3

# set relative path to the database
PATH_TO_DB = "./sqlite/musicians.db"
```

Next, we create a *connection* to the database, and we create a *cursor*. The cursor will allow for defining a set of rows and performing complex logic on them.

```
# create db connection
conn = sqlite3.connect(PATH_TO_DB)

# create cursor
cur = conn.cursor()
```

Next, we define some artist data. For this small example, we create a list of tuples containing artist name and genres. For example, Taylor Swift's music can be categorized as country and pop.

```
artists = [
        ('Taylor Swift', "['country', 'pop']"),
        ('Chris Stapleton', "['country','soul','rock',
                              'bluegrass']"),
        ('Bono Hewson', "['rock','pop']")]
```

To create the table in the database, we will need to provide a schema. For each field, the schema needs to provide:

- The field name

- The field data type

- Optional attributes, such as whether the field may contain null values

Each database *transaction* ends with a commit command. Let's create the table, passing the schema:

```
query = 'CREATE TABLE artist (artist_name string, genre string);'
cur.execute(query)
conn.commit()
```

Next, insert multiple records of data with executemany(). For each record, the placeholder (?,?) will be populated with the two columns of data in artists. The INSERT INTO statement is how records are written to database tables with SQL.

```
query = 'INSERT INTO artist VALUES (?,?);'
cur.executemany(query, artists)
conn.commit()
```

Let's verify that the data was stored properly. We will use a SELECT statement
to select all rows and columns. The '*' token will fetch all columns.

```
query = 'SELECT * FROM artist;'
for row in conn.execute(query):
    print(row)
```

```
OUT:
('Taylor Swift', "['country','pop']")
('Chris Stapleton', "['country','soul','rock','bluegrass']")
('Bono Hewson', "['rock','pop']")
```

Note that the loop for fetching rows is required by the API. If we wanted to
fetch all of the rows in SQLite at the command line, we would simply run the
query:

```
sqlite> select * from artist;
```

```
OUT:
Taylor Swift|['country','pop']
Chris Stapleton|['country','soul','rock','bluegrass']
Bono Hewson|['rock','pop']
sqlite>
```

The result set has a different syntax, but it contains the same data.

Next, let's select only the musicians who sing country music. Filtering is done
with the WHERE command. The text match can be implemented with the LIKE
command like this:

```
query = "SELECT * \
        FROM artist \
        WHERE genre LIKE '%country%';"
```

```
for row in conn.execute(query):
    print(row)
```

```
# NOTES:
# The character '\' is a line break for readability
# The character '%' is a wildcard for string matching
# This will match on genres containing 'country'
```

```
OUT:
('Taylor Swift', "['country','pop']")
('Chris Stapleton', "['country','soul','rock','bluegrass']")
```

This gives the correct output, but we really only want the artist names. We were selecting all columns with '*', but we can specify the columns that we want. Let's run the same query but SELECT only the artist_name column:

```
query = "SELECT artist_name \
        FROM artist \
        WHERE genre LIKE '%country%';"

for row in conn.execute(query):
    print(row)
```

OUT:
```
('Taylor Swift',)
('Chris Stapleton',)
```

This has the right output. If we would like the artist names in a list, we can store the first element of each tuple like this:

```
data = []
query = "SELECT artist_name \
        FROM artist \
        WHERE genre LIKE '%country%';"

for row in conn.execute(query):

    # extract artist_name and append to list
    data.append(row[0])

print(data)
```

OUT:
```
['Taylor Swift', 'Chris Stapleton']
```

Storing the result set allows for downstream use later.

Counting and Aggregating

It is often useful to count the number of records in a table to understand its size. This can be done as follows:

```
query = 'SELECT count(*) FROM artist;'
conn.execute(query).fetchone()[0]
```

OUT:
```
3
```

The `count(*)` function counts the number of records. The `fetchone()` function retrieves the result set.

Data aggregations are useful for partitioning a dataset and computing statistics on each piece of the partition. An example is the calculation of a town's daily average rainfall by month. These calculations are commonly and efficiently performed with a *Split-Apply-Combine* pattern. The first step in the strategy is to define one or more grouping variables (in this case, the grouping variable would be *month*). Next, the records are split into discrete groups. For the records in each group, a function is applied to the values. Here, we calculate the mean rainfall over the days in January, February, and so on. This yields the daily average rainfall for each month. In the final step, the results from each group (the months) are combined into a data structure, such as a table.

The Split-Apply-Combine pattern is supported in SQL and many other languages including Python and R. Let's demonstrate how to aggregate data using SQL. We begin by creating a small dataset of concert venues with their states:

```
query = 'CREATE TABLE venues
          (venue_name string, venue_state string);'

cur.execute(query)
conn.commit()

venue_data = [('Red Rocks', 'CO'),
              ('Santa Barbara Bowl', 'CA'),
              ('Greek Theater', 'CA'),
              ('Madison Square Garden', 'NY')]

query = 'INSERT INTO venues VALUES (?,?);'
cur.executemany(query, venue_data)
conn.commit()

query = 'SELECT * FROM venues;'
for row in conn.execute(query):
    print(row)

OUT:
('Red Rocks', 'CO')
('Santa Barbara Bowl', 'CA')
('Greek Theater', 'CA')
('Madison Square Garden', 'NY')
```

Now we would like to calculate the number of concert venues in each state. This can be done by:

- Forming groups of records by the grouping variable `venue_state`

- Counting the number of records in each group

- Combining the counts per state

The aggregation can be done like this:

```
query = 'SELECT venue_state, count(*) \
        FROM venues \
        GROUP BY venue_state;'

for row in conn.execute(query):
    print(row)
```

```
OUT:
('CA', 2)
('CO', 1)
('NY', 1)
```

The line of code

```
SELECT venue_state, count(*)
```

will return each state and its associated count.

The line

```
GROUP BY venue_state
```

will perform the grouping operation.

From the result set, there are tuples containing each state and its associated number of venues. The `GROUP BY` command is very powerful and can aggregate across multiple grouping variables.

In the next section, we will get to the main purpose of relational databases: joining datasets. This will allow for queries which can extract information from across the database.

7.4 Music Query: Multiple Tables

When doing data science, the required data may be spread across two or more tables. In such a case, we can join the tables two at a time on common fields.

Joins bring a bit of complexity, so we will first study some join examples. Then we will continue to build out the music artist example.

Joins

Common joins are `INNER`, `OUTER`, `LEFT`, and `RIGHT`. These options provide flexibility for joining tables with missing data. When joining two tables L and R, we call L the left table and R the right table. Let's look at some sample data to understand how the different joins work.

Table: L

id	first_name	favorite_pet
0	Taylor	rabbit
1	Chris	cat
2	Bono	dog

Table: R

id	first_name	favorite_number
1	Chris	314
2	Bono	5
3	Cher	12

Suppose we join tables L and R on the `id` field, selecting all of the unique columns. Both tables have records with `id` 1 and 2; we say these records match. However, `id=0` is only in table L, while `id=3` is only in table R. The type of join we should use will depend on which records we wish to keep.

A `LEFT` join will return all of the records from the left table, and only the matching records from the right table. The result looks like this:

LEFT JOIN

id	first_name	favorite_pet	favorite_number
0	Taylor	rabbit	None
1	Chris	cat	314
2	Bono	dog	5

Since we used a `LEFT` join, only the records from table L were included, and so Cher was excluded. Since Taylor is not in table R, we don't have her favorite number.

A `RIGHT` join will return all of the records from the right table, and only the matching records from the left table. The result looks like this:

RIGHT JOIN

id	first_name	favorite_pet	favorite_number
1	Chris	cat	314
2	Bono	dog	5
3	Cher	None	12

Since we used a `RIGHT` join, only the records from table R were included, and so Taylor was excluded. Since Cher is not in table L, we don't have her favorite pet.

An `INNER` join will return only the matching records from both tables. This means that any non-matching records will be excluded in the join. The result looks like this:

INNER JOIN

id	first_name	favorite_pet	favorite_number
1	Chris	cat	314
2	Bono	dog	5

Inner joins produce simpler datasets as there are no missing values to handle. However, this means that records may be dropped. Sorry Taylor and Cher!

An `OUTER` join will return all records from the tables. The result looks like this:

OUTER JOIN

id	first_name	favorite_pet	favorite_number
0	Taylor	rabbit	None
1	Chris	cat	314
2	Bono	dog	5
3	Cher	None	12

The outer join may produce records with missing data, as in this case.

Joining the Musician Tables

Next, we return to our music artist data. We will create a table in the database called `hometown` and pass the schema. This will hold the hometowns of the artists.

```
query = 'CREATE TABLE hometown \
            (artist_name string, hometown_name string);'

cur.execute(query)
conn.commit()
```

Creating some data for the hometown table:

```
data_hometown = [
                ('Taylor Swift', 'West Reading, Pennsylvania'),
                ('Chris Stapleton', 'Lexington, Kentucky'),
                ('Bono Hewson', 'Dublin, Ireland'),
                ('Rihanna', 'Saint Michael, Barbados')
                ]
```

Notice that Rihanna is in this dataset, but not in the artist table. This will have implications in the joins that we demo later. The following code will insert the hometown records into the table:

```
query = 'INSERT INTO hometown VALUES (?,?);'
cur.executemany(query, data_hometown)
conn.commit()
```

Next, we query the table to verify the data has been loaded:

```
query = 'SELECT * FROM hometown;'
for row in conn.execute(query):
    print(row)
```

```
OUT:
('Taylor Swift', 'West Reading, Pennsylvania')
('Chris Stapleton', 'Lexington, Kentucky')
('Bono Hewson', 'Dublin, Ireland')
('Rihanna', 'Saint Michael, Barbados')
```

This is correct. Next, let's try out some joins on the hometown and artist tables. We can use the common field artist_name for the join. First, we run an INNER JOIN:

```
query = 'SELECT hometown.artist_name, \
        hometown.hometown_name, \
        artist.genre \
        FROM hometown INNER JOIN artist \
        ON hometown.artist_name = artist.artist_name;'

for row in conn.execute(query):
    print(row)
```

```
OUT:
('Taylor Swift',
 'West Reading, Pennsylvania',
 "['country','pop']")
```

```
('Chris Stapleton',
 'Lexington, Kentucky',
 "['country','soul','rock','bluegrass']")

('Bono Hewson',
 'Dublin, Ireland',
 "['rock','pop']")
```

Three records came back from the query, and they are formatted for readability. Since we did an INNER JOIN and Rihanna was not in the `artist` table, she was dropped from the join. The other three artists were in both tables and were retained.

There is some new syntax: a query with multiple tables needs to request fields with their table names, as in `hometown.artist_name`. This makes things clearer, and avoids a collision when the same field is in both tables. The line:

```
FROM hometown INNER JOIN artist
```

specifies the tables and the join type.

The line:

```
ON hometown.artist_name = artist.artist_name
```

matches the records on the common field.

Next, let's run a LEFT JOIN. This can be accomplished with a small code change: replace INNER JOIN with LEFT JOIN.

```
query = 'SELECT hometown.artist_name, \
         hometown.hometown_name, \
         artist.genre \
         FROM hometown LEFT JOIN artist \
         ON hometown.artist_name = artist.artist_name;'

for row in conn.execute(query):
    print(row)

OUT:
('Taylor Swift',
 'West Reading, Pennsylvania',
 "['country','pop']")

('Chris Stapleton',
 'Lexington, Kentucky',
 "['country','soul','rock','bluegrass']")
```

```
('Bono Hewson',
 'Dublin, Ireland',
 "['rock','pop']")

('Rihanna',
  'Saint Michael, Barbados',
  None)
```

Since we are treating `hometown` as the left table, we get back each record from this table. Since Rihanna was not in the `artist` table, we don't have the associated data for `genre`, and it appears as None.

7.5   Houses, Lakes, and Lake Houses

As organizations grow, the quantity of data tends to grow with it. This can be driven by factors including more products, more teams, and more customers. A single database often won't be sufficient. A data strategy that will support the business should take these data storage requirements into account:

- Variety: The data may come from a variety of sources and it may have different levels of structure. There may be a need to collect transactional data, which is very structured, as well as documents, images, and videos, which are unstructured.

- Volume: There may be a massive amount of data to store and retrieve. It is important to support these requirements at the necessary scale.

- Access: Different teams and users will have different permissions. A team that performs analytics on home equity loans might not need access to student loan data. A business intelligence team might need to compute analytics across all of the lines of business. An ideal system would make it easy for an administrator to grant and monitor access.

- Cataloging: It is valuable for permissioned users to browse datasets at a high level, and request access to what they need. Metadata saved with the datasets can help users find what they need.

There may be many other organizational requirements as well. Even with this short list, however, a single database likely won't be sufficient. In response, several objects and concrete services have emerged. A very brief overview follows next, but bear in mind that each of these topics could be their own book. Let's start with the data warehouse.

7.5.1 Data Warehouse

A data warehouse is a central place for data storage. It integrates data from one or more sources, which may include transactional systems and relational databases. The implications are that data is copied, and processes are required for preparing and storing the data. The data warehouse stores structured data only. The ETL process of extracting source data, transforming it, and loading it into the warehouse generally runs on a regular schedule.

The purpose of the data warehouse is to allow for enterprise-wide reporting, ad hoc queries, data analysis, and decision making. This is a key function in *business intelligence* (BI). A business intelligence analyst will use SQL heavily. While BI and data science often have different mandates and the participants are on separate teams, there can be fruitful interaction. In particular, BI teams are generally highly knowledgeable about the availability, lineage, and readiness of the datasets across the firm. The right datasets can enable broad analyses, such as calculating the profitability across all lines of business.

Several commercial, cloud-based data warehouse products exist on the market, such as Amazon Redshift, Google BigQuery, and Snowflake. Working in these systems is similar to working in a RDBMS user interface. Panes are available for essential functions like browsing objects (e.g., tables and views), working with SQL code, and reviewing result sets.

7.5.2 Data Lake

Given the variety of modern data sources and formats, a storage solution is needed for semi-structured and unstructured data. For some files like images and videos, the data cannot be stored in tables. For other files like documents, it may be preferable to store the data in its native format first and figure out a proper schema later. This is because it can take a lot of time to understand the important pieces of data and their relationships to one another. Finally, even if a schema is specified, it might evolve, and a tabular format might not be the best structure. Requiring the schema upfront is too strict a requirement for some data.

The *data lake* allows for storing data in a variety of formats without specifying a schema. Like the data warehouse, the data lake can be used as a central place for storage. Unlike the data warehouse, ETL jobs are not needed for extracting and preparing the data for storage in the lake. For data that is cleaned and prepared in the data lake, it can be stored in distinct zones for easy identification and retrieval (think of these as separate folders).

Some of the early data lakes were built by companies and managed on-premise. This generally required a team of IT specialists to manage software, hardware, climate control, and so forth. As the stockpile of data outgrew the on-premise servers, new, costly machines needed to be added. In short, it was a resource-intensive endeavor that didn't scale well and often didn't align with the value proposition of the organization (e.g., does a pharmaceutical company need to be doing this?). Today, it is generally more economical for organizations to outsource this function, which means using one of the data lakes offered by a cloud provider. As mentioned earlier, examples of data lakes include Amazon S3 (Simple Storage Service), Azure Data Lake, and Google Cloud Storage.

While it is easy to quickly ingest, store, and centralize all manner of data in the data lake, a new problem arises: the lake can become polluted with improper and incorrect data. The popular term for this phenomenon is the data lake becoming a *data swamp*. The fear is that the data swamp will render the data lake less useful, and this is a real concern.

7.5.3 Data Lakehouse

The data lakehouse is a hybrid of a data lake and a data warehouse. It offers a centralized data store for data of any structure with the ability to save tabular forms as needed. In this way, users can impose structure on segments of the data, leveraging the strength of a data warehouse. This helps to minimize the risk of the data storage becoming a swamp.

In addition to the data lakehouse scaling well to a variety of datasets, a robust offering will provide functionality for building a data catalog. This allows users to easily browse available datasets, which can improve projects and encourage new ones. Finally, administrators can manage access to users, which promotes secure, efficient data sharing.

The data lakehouse is frequently evangelized as the modern data architecture. Several cloud providers offer commercial data lakehouse products, including Databricks and AWS.

7.6 Chapter Summary

A database supports the creation, retrieval, updating, and deletion of data, or the CRUD operations. Relational databases can be used to separate concepts into tables to maximize efficiency and minimize duplication. The tables can be joined as needed to bring data together. SQL is used to work with relational databases. Several relational database systems (RDBMS) are on the market, and some are freely available. Non-relational databases, or NoSQL databases, support semi-structured and unstructured data.

A table consists of rows and columns. Rows are also called records and tuples, while columns are also called attributes and fields. The primary key is the unique identifier for telling records apart; it may consist of multiple fields. A foreign key is a field in one table that references the primary key in another table.

We reviewed some SQL commands for creating tables, inserting data, retrieving data, counting and aggregating, and joining tables. We used SQLite inside Python with the help of an API. The API includes functions for working with SQL, and so the code is a little different from writing straight SQL. We discussed the importance of data scientists knowing SQL well, as it will be used regularly.

For a complete storage solution at an organization, one database generally won't be sufficient. A robust storage strategy will scale, handle a variety of data, and allow easy, transparent data cataloging and access. A data warehouse is a central place for integrating data from multiple sources. It only stores structured data, which limits its effectiveness. A data lake can be a central data store for semi-structured and unstructured data. It can scale, but its quality can be compromised since there is no structure imposed. A data lakehouse is a hybrid of a data lake and a data warehouse. It offers the ability to scale and store structured data. It is frequently touted as the way forward in data storage solutions.

In the next chapter, we will spend some time building our mathematical toolbox. From there, we will build our statistical knowledge up from our probability foundation. This will make our analytical capabilities more robust.

Going Further

Many data storage tools were introduced in this section. Here are some starting points for further investigation:

An outline and comparison of the database, data warehouse, and data lake: `https://aws.amazon.com/data-warehouse/`

Key-value databases: `https://redis.com/nosql/key-value-databases/`

Graph databases: `https://neo4j.com/developer/graph-database/`

Document databases: `https://www.mongodb.com/document-databases`

7.7 Exercises

Exercises with solutions are marked with ⑤. Solutions can be found in the book_materials folder of the course repo at this location:

https://github.com/PredictioNN/intro_data_science_course/

1. Ⓢ Can an Excel spreadsheet be considered a database? Explain your answer.

2. Ⓢ True or False: For any relational database table, there needs to be one field which has unique values.

3. Ⓢ Rows in a database table are also called x and y. Columns in a database table are also called a and b. What are x, y, a, and b?

4. Ⓢ A database administrator wishes to prevent users from inserting an invalid field value into a table. What can be done to support this requirement?

5. Ⓢ Explain the difference between a primary key and a foreign key.

6. Ⓢ A relational database uses multiple tables. Explain at least one benefit to this approach.

7. When should a relational database be used? When should it not be used?

8. For the `artist` table, write and execute an `SQL` query that selects the musicians who sing pop music.

9. For the `venues` table, insert two additional concert venues with their names and states. Next, calculate the number of venues by state, and sort descending by the counts. That is, the state(s) with the fewest venues should be listed first.

10. A user would like to join two database tables. What is required to do this?

11. Ⓢ True or False: An outer join will merge two tables and produce a new table with no missing values.

12. How is an inner join different from a left join?

13. Compare and contrast a database and a data warehouse.

14. Ⓢ When is it preferable to use a data lake instead of a data warehouse?

15. Ⓢ Discuss a large challenge faced by data lakes.

16. Ⓢ Read about *eventual consistency* and summarize the premise in a few sentences. What is the benefit to eventual consistency?

8

Mathematics Preliminaries

8.1 Set Theory

Set

A *set* is a collection of objects. Sets arise often in data science, for example when we consider the set of unique users, the set of videos they've watched, or the set of Boolean predictors in a dataframe.

Additional sets include:

- The set of integers $\mathbb{Z}$

- The set of real numbers $\mathbb{R}$

- The set of positive real numbers $\mathbb{R}^+ = \{x \in \mathbb{R} : x > 0\}$

Set Builder Notation

In the last case, *set builder notation* denotes each element from the real numbers ($x \in \mathbb{R}$) such that the element is greater than zero. This notation is useful when it is tedious or not possible to enumerate all elements in a set.

Subset

For sets A and B, we say A is a *subset* of B (written $A \subseteq B$) if every element of A is also an element of B. This allows for the possibility that A and B contain exactly the same elements, in which case $A = B$. In the case where every element of A is an element of B and they are not equal (for example, B contains additional elements), then A is a *proper subset* of B, written $A \subset B$.

Empty Set

The *empty set*, denoted $\{\}$ or $\emptyset$, contains no elements and is a subset of every set. This set plays a prominent role, for example when testing if a filter returns any elements from a dataset.

Cardinality

The size or *cardinality* of a set A denoted $|A|$ is the number of elements in the set. Here is a small code example of creating an empty set and computing its cardinality:

```
empty = {}
print(empty)
print(len(empty))
```

OUT:
```
{}
0
```

Set Operations

For a *universal set* of all elements U with subsets A and B, it will be useful to find:

- The *complement* of A, which is a set containing elements in U but not in A, denoted A^c

- The set of elements in A and B; this is their *intersection*, $A \cap B$

- The set of elements in A, B, or both; this is their *union*, $A \cup B$

- The set of elements in A but not in B; this is their *set difference*, $A - B$

Here is a small example:

$U = \{a, b, c, d, e, f, g\}$
$X = \{a, c, e\}$
$Y = \{a, c\}$
$Z = \{b, c, d\}$

Based on these sets, Y is a proper subset of X since Y contains all elements in X but they are not equal. This is denoted $Y \subset X$.

$X^c = \{b, d, f, g\}$.

$X \cap Z = \{c\}$

$X \cup Z = \{a, b, c, d, e\}$

$X - Y = \{e\}$

We can determine these sets in Python like this:

```
U = {'a','b','c','d','e','f','g'}
X = {'a','c','e'}
Y = {'a','c'}
Z = {'b','c','d'}
```

```
print('X_complement:',  U - X)
print('X intersection Z:',  X.intersection(Z))
print('X union Z:',  X.union(Z))
print('X minus Y:', X - Y)

OUT:
X_complement: {'g', 'b', 'f', 'd'}
X intersection Z: {'c'}
X union Z: {'b', 'd', 'e', 'c', 'a'}
X minus Y: {'e'}
```

Note that to determine the complement of X, it was necessary to compute it relative to the universal set U.

It will be useful to check if an element is in a set. Here are code examples:

```
print('g' in Z)
print('g' not in Z)
print('b' in Z)
print('b' not in Z)

OUT:
False
True
True
False
```

The results follow from the logic of g not belonging to Z, and b belonging to Z.

8.2 Functions

For sets A and B, a *function* is a rule or *mapping* that takes an element of A as input and assigns an element of B as output. This is commonly written $f : A \to B$ for a function f. The space of inputs A is called the *domain*, and the space of outputs B is called the *range*. The value of the function at a point x can be written $f(x)$. Note that for each input value, there can only be one output value. The *vertical line test* is a graphical method for checking if a curve is a function. If a vertical line intersects the curve in more than one place, then it is not a function. Note that we might define a function using the y notation as in $y = 2x$ or using the $f(x)$ notation as in $f(x) = 2x$ to clarify the functional dependence on x.

It is completely valid for a function to map two or more different inputs to the same output. Figure 8.1 shows the graph of the function $f(x) = x^2$. Both $-x$ and x will map to the same value $f(-x) = f(x)$ for all x. For example, $f(-2) = f(2) = 4$.

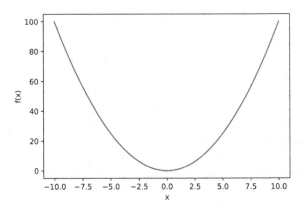

FIGURE 8.1: The function $f(x) = x^2$

Function Composition

Function composition is an operation $\circ$ that takes two functions f and g and produces a new function $h = g \circ f$ such that $h(x) = g(f(x))$. The function f is applied, followed by the function g. For cases where more than two functions are composed, say $k \circ g \circ f$, we apply the functions from right to left, taking them two at a time. For example, $k \circ g \circ f = k(g(f(x))) = k(h(x))$, by applying the equation $h(x) = g(f(x))$. The evaluation of this nested composition is supported by the associative property of function composition.

Consider an example with two functions $f(x) = x + 1$ and $g(x) = \sqrt{x}$, each defined for the set of real numbers. We form the function composition $h(x) = g(f(x)) = \sqrt{x + 1}$. Can we compute $h(3)$ and $h(-2)$? The first case is possible, with $h(3) = \sqrt{3 + 1} = \sqrt{4} = 2$. The second case is undefined, since $h(-2) = \sqrt{(-2) + 1} = \sqrt{-1}$. Hence, composition will only work where the domain of h is defined. In this case, the domain is $x \geq -1$.

Function composition is particularly important in data science because many machine learning models apply a sequence of functions to data. In computer vision, image pixels can be arranged in layers and then aggregated and averaged many times to infer whether an object of interest can be found.

Function Inverse

The *inverse function* of a function f serves to undo the function. It is itself a function. If $f(x) = y$, then entering y into the inverse function f^{-1} produces the output x. This can be written $f^{-1}(y) = x$. We can use function composition to show that $f^{-1}(f(x)) = f^{-1}(y) = x$. Not every function has an inverse, so we will discuss the necessary conditions next. A function that has an inverse is called *invertible*.

Does our parabola function from Figure 8.1 have an inverse? A function inverse is itself a function, which means that each input must map to one output. Remember that the inverse works backwards, so y is treated as input and x is treated as output. We want to confirm that each possible y value maps to only one x.

Let's select the value 60 on the y-axis and check if it maps to one value on the x-axis. We can draw a horizontal line at 60, check where the line intersects the graph, and drop vertical lines to the x-axis at those intersections. It turns out there are two x-values meeting this condition: $-\sqrt{60}$ and $\sqrt{60}$. In fact, due to symmetry, every y-value except the lowest point will return two values. Since we cannot uniquely identify an x for any $f^{-1}(y)$, the inverse does not exist. This *horizontal line test* can be a useful way to check for a function inverse.

Our quick test included two criteria, so let's expand on them. First, a function f is one-to-one if $f(x) \neq f(y)$ whenever $x \neq y$. That is, if two inputs are different, their outputs must be different. The parabola violated this rule, as $f(\sqrt{60}) = f(-\sqrt{60}) = 60$. Second, a function f is *onto* if for each $b \in B$ there is at least one $a \in A$ with $f(a) = b$. For illustration, imagine that A is a set of cat lovers and B is a set of cats, and there are more cat lovers than cats ($|A| > |B|$). An animal shelter asks each cat lover to indicate their top candidate for adoption, imposing a strict rule that each cat must be adopted. This creates an onto function, whereby each element of B (the cats) is matched with at least one element of A (the cat lovers).

Finally, a function f has an inverse f^{-1} if it is one-to-one and onto. This inverse is unique.

Monotonicity

Some functions have the interesting property that they never increase or never decrease as x increases. A function that increases or remains constant as x increases is called a *monotonically increasing function*, as in Figure 8.2. A function that decreases or remains constant as x increases is called a *monotonically decreasing function*.

A function is *strictly increasing* if it always increases as x increases, and it is *strictly decreasing* if it always decreases as x increases. Functions meeting either of these conditions are *strictly monotone*.

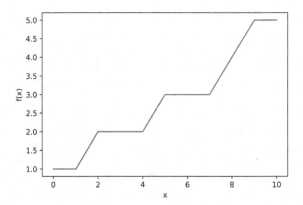

FIGURE 8.2: A Monotonically Increasing Function

There are many practical uses of monotonicity in data science including:

- Finding the maximum or minimum (the *extrema*) of a function over a domain. If we know the function monotonically increases in a linear fashion, for example, then the minimum must occur for the smallest value in the domain.

- Simplifying computation. Sometimes the order of values is important (e.g., $x_1 > x_2 > \ldots > x_n$) but the values themselves $x_1, x_2, \ldots, x_n$ are not important. Applying a function that preserves the order can ease computational burden. A common choice is $log(x)$, which produces $(log(x_1), log(x_2), \ldots, log(x_n))$ in this example.

Convexity

Loosely speaking, a function f is *convex* if it "forms a cup." Geometrically, for a convex f, any line segment connecting two points on the graph of f would lie above the graph between the two points. Figure 8.3 provides an illustration.

The convexity property of a function is very useful when finding its lowest value on a domain, or *global minimum*. If an algorithm visits a sequence $x_1, x_2, x_3, \ldots$ which maps to decreasing function values $f(x_1), f(x_2), f(x_3), \ldots$, it will eventually reach the function minimum. For a non-convex function, reaching the global minimum may be difficult or even impossible. Figure 8.4 shows a graph with many *local minima* – which are minima over subsets of the domain – and a single global minimum near the middle of the domain. Given the local minima, it becomes much harder for an algorithm to find the global minimum. Specifically, success will depend on where the algorithm starts the search, and it would need to start near the global minimum.

If a function $-g$ is convex, then g is a *concave* function. Loosely speaking, a concave function "forms a hill." Concave functions are useful when finding

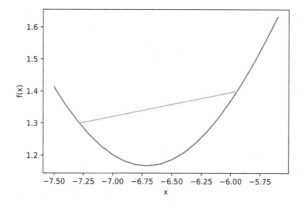

FIGURE 8.3: A Convex Function

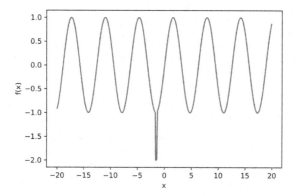

FIGURE 8.4: A Challenging Global Minimum

their highest value on a domain, or *global maximum*. Analogous to convex functions, it can be difficult or even impossible to find a global maximum on a non-concave function with many local maxima.

Later in the book, we will fit models to data. The models will depend critically on parameters, because a poor choice of parameter values will produce predictions with greater deviations from the target values. To determine the best parameter values, a function is selected to represent a measure of error (to be precisely defined later). We will use convex functions to more easily determine parameter values minimizing this error.

A Summary of Function Types

Next we discuss different types of functions: *rational functions*, *algebraic functions*, *trigonometric functions*, *exponential function*, and the logarithm. Each of these function types is useful in data science.

Rational integral functions, more commonly called *polynomials*, are the simplest functions. They are obtained by repeated application of elementary operations: addition, subtraction, and multiplication. Given constants $a_0, a_1, \ldots, a_n$, polynomials can be represented like this:

$$y = a_0 + a_1 x + \ldots + a_n x^n$$

Examples include $y = 3x+4$ and $y = 1+x+x^2$; the former is a *linear function* and the latter is a *quadratic function*.

A rational function is a fraction with polynomials in the numerator and denominator. The denominator cannot be zero. Examples include $y = \frac{1}{1-x}$ and $y = \frac{x}{1+x^2}$.

Algebraic functions can be defined as the root of a polynomial equation. For illustration, let's start with $y = x^n$, a polynomial. We will require $x \geq 0$ and find the inverse of the function. This can be done by solving for x. We can then interchange x and y, which produces $y = x^{1/n} = \sqrt[n]{x}$. The case of $n = 2$ produces the rational function $y = \sqrt{x}$, for example.

The *elementary transcendental functions* – comprising the trigonometric functions, the exponential function, and the logarithm – derive from geometry. Consider the circle of radius 1, or *unit circle*, in Figure 8.5. The angle θ is measured in *radians*, which measures the length of the arc of the circumference that the angle sweeps out. For the unit circle, the circumference is $C = 2\pi r = 2\pi$. This means that an angle sweeping out the entire circumference, or 360 degrees, would measure 2π, while an angle sweeping out half the circle, or 180 degrees, would measure π.

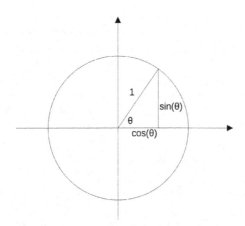

FIGURE 8.5: Unit Circle

Notice the right triangle formed in the figure. The horizontal side formed from the x-axis is called the *adjacent* side, the vertical side dropped from the circle to the x-axis is called the *opposite* side, and the side forming the radius is called the *hypotenuse*. The trigonometric functions *sine*, *cosine*, and *tangent* are defined as ratios of these triangle sides:

$$\sin(\theta) = \text{opposite/hypotenuse}$$
$$\cos(\theta) = \text{adjacent/hypotenuse}$$
$$\tan(\theta) = \sin(\theta)/\cos(\theta) = \text{opposite/adjacent}$$

Since the hypotenuse has length 1 for the unit circle, it follows in this example that the adjacent side is equal to $\cos(\theta)$ and the opposite side is equal to $\sin(\theta)$. Figure 8.6 shows the periodic nature of the sine and cosine functions, which is useful for modeling cyclical data (e.g., sales cycles, seasonal variation). Their patterns are identical but shifted by a fixed quantity.

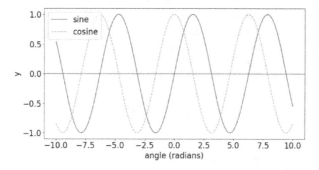

FIGURE 8.6: Sine and Cosine Functions

The exponential function with base a (for any $a > 0$) and rational number[1] x can be written $y = a^x$. The inverse of the exponential function is the logarithm with base a, which can be written $x = log_a(y)$. The logarithm can be determined by answering the question "the base to *which* power is y?" For example, in solving $log_2(8)$, we ask: 2 to which power is 8? The solution is 3.

A very special base is the number e, which can be understood like this:

Compute the sum $S_n = 1 + \frac{1}{1!} + \frac{1}{2!} + \ldots + \frac{1}{n!}$ for some n.

The expression $n!$ is read "n factorial" and computed as $n \times (n-1) \ldots \times 1$. For example, $3! = 3 \times 2 \times 1 = 6$. As n increases, the sum will approach, or *converge to*, a number less than 3. This number is an irrational number with approximate value 2.71828. Let's look at a few of the sums:

---

[1]Rational numbers can be written as a fraction, such as 2 or 2/3.

$$S_2 = 1 + \frac{1}{1!} + \frac{1}{2!} = 2.5$$

$$S_3 = 1 + \frac{1}{1!} + \frac{1}{2!} + \frac{1}{3!} \approx 2.67$$

$$S_4 = 1 + \frac{1}{1!} + \frac{1}{2!} + \frac{1}{3!} + \frac{1}{4!} \approx 2.71$$

More formally, e can be defined with a statement using a *limit*, where the limit will be formally defined in the next section on differential calculus.

$$e = \lim_{n \to \infty} S_n$$

The exponential function $y = e^x$ is very common and it appears in probability functions that we will soon encounter. The inverse of this function is a logarithmic function; it is given the special notation $ln(x)$ which denotes the *natural logarithm*. The base is generally not written, as it is understood to be e. For example, $ln(e^3) = 3$, as e to the power of 3 is e^3. A second example is $ln(1) = 0$, as e to the power of 0 is 1.

Figure 8.7 shows the functions e^x and $ln(x)$. As x increases, e^x increases to $+\infty$. As x decreases, e^x tends to zero (it approaches an *asymptote* at zero). Its domain is therefore $(-\infty, +\infty)$, while its range is $(0, +\infty)$. The function $ln(x)$ increases as x increases, but the rise is much slower than e^x. As x decreases to zero, $ln(x)$ approaches $-\infty$. The domain is $(0, +\infty)$, while the range is $(-\infty, +\infty)$. As e^x and $ln(x)$ are inverses, their domains and ranges are reversed.

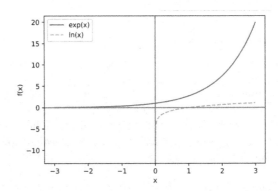

FIGURE 8.7: Exponential and Natural Logarithm

Vectorized Functions

It is often useful to apply a function to several values at once, and this is supported by *vectorization*. Here is an example of using numpy to compute the square root of a list of numbers:

```
x=[1,4,9,16]
np.sqrt(x)
```

```
OUT:
array([1., 2., 3., 4.])
```

In this case, the function was applied to each element in the list: $(\sqrt{1}, \sqrt{4}, \sqrt{9}, \sqrt{16})$. Note that not every module supports vectorization. If we attempt this with the `math` package, it emits an error:

```
x=[1,4,9,16]
math.sqrt(x)
```

```
OUT:
TypeError: must be real number, not list
```

The error indicates that `sqrt()` from the `math` package should be applied to a single value.

8.3 Differential Calculus

Data science uses models which rely on important properties like the continuity of a function. This is treated with calculus, which belongs to the branch of mathematics called *analysis*. Next, we will take a quick tour to learn some essential calculus concepts.

Limit

The concept of a limit, mentioned earlier, is fundamental to analysis. The limit of a sequence can be explained with an example:

Consider the sequence with terms:

$$a_1 = 1, \quad a_2 = \frac{1}{2}, \quad a_3 = \frac{1}{3}, \quad \ldots, \quad a_n = \frac{1}{n}, \quad \ldots$$

As we look at terms deeper into the sequence (with larger n), they get closer to zero. None of the terms will be equal to zero, but we can get arbitrarily close to zero. In fact, there is a term a_N such that all subsequent terms are sufficiently close to zero. For example, if we wish to be within $1/10th$ of zero, then terms beyond $a_{10} = \frac{1}{10}$ will suffice. We can express this notion by saying that as n increases, the terms a_n tend to zero, or that they have a limit of 0. We can also say that the sequence converges to zero, denoted:

$$\lim_{n \to \infty} a_n = 0$$

A sequence *diverges* if it does not converge to any number.

Next we turn from the limit of a sequence to the limit of a function. The central idea is the same: getting arbitrarily close to some value (the limit) by changing the input value. For the limit of a sequence, we can get arbitrarily close by sufficiently increasing n. For the limit L of a function $f(x)$, we can get arbitrarily close by moving x closer to some point ξ. We say that the value of $f(x)$ tends to limit L as x tends to ξ, which can be written symbolically:

$$\lim_{x \to \xi} f(x) = L$$

Returning to the exponential function $y = e^x$, we noticed that as x decreased, e^x approached zero. In fact, this is an example of a limit, which can be expressed:

$$\lim_{x \to -\infty} e^x = 0$$

What happens to $y = e^x$ as x increases? As an exponential function, the value increases rapidly and without bound. For any threshold, the function would eventually cross it for sufficiently large x. This is an example of divergence; as x increases, the limit does not exist. This can be written:

$$\lim_{x \to \infty} e^x = \infty$$

Continuity

A function is *continuous* on an interval if it can be "drawn without taking pencil off the paper." A continuous function has no breaks or jumps. Consider a function that has value 0 for all $x < 0$ and value 1 for all $x \geq 1$. Revisiting limits, we would say that the function has limit 0 when approaching x from the left of 0, and limit 1 when approaching from the right of zero. These two limits do not coincide due to a break, or *jump discontinuity*. We say the function has a *discontinuity* at this point, and it is *discontinuous* on an interval containing 0.

Practically speaking, it is often important to work with functions that are continuous, as there is a guarantee that we can get arbitrarily close to a point $f(x_0)$ by moving x close to some point x_0. If it were not possible to get arbitrarily close, this would imply that changing x by some small amount might change $f(x)$ by a large amount (as in the case of the jump discontinuity). Imagine gently holding the button on a remote control to increase the volume. We expect the volume to increase steadily, and not suddenly jump to a deafening level.

We can state this mathematically as follows:

A function $f(x)$ is continuous at a point x_0 if for every positive number ϵ, there exists another positive number δ such that $|f(x) - f(x_0)| < \epsilon$ for all x for which $|x - x_0| < \delta$.

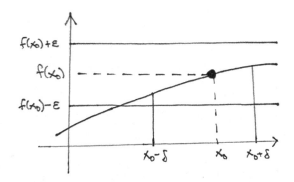

FIGURE 8.8: Continuity at a Point

We have defined continuity at a point. For a function to be continuous on an interval, it must be continuous at each point in the interval.

Derivative

We often use functions to model behavior: given input x, what is output $f(x)$? Another important question is about sensitivity to the input: when we change x by some "small" amount, what will be the change to $f(x)$? If we press down more firmly on the car's accelerator, what will be the change in position over a given time increment? The *derivative* measures this change in output with respect to a change in input.

We start with a discussion of the derivative of a single-variable function, and then to the case of a function of multiple variables. In the former case, the derivative at a given value x is the slope of the *tangent line* to the graph of the function at that point. Figure 8.9 shows an example. At the point $x = 1.5$, the derivative is approximately 0.408 (we soon discuss how to calculate this), which is equal to the slope of the line which is tangent to the function at this point. For the tangent (or derivative) to exist at this point, the function must be continuous at this point. If we move the point along the curve, then the derivative will change in accordance with the behavior of the function.

Let's quickly determine the derivative of some simple functions. The constant function $f(x) = 1$ is a horizontal line, and the slope of a horizontal line is zero. Hence, the derivative is zero everywhere. As a second case, the line $f(x) = 2x$ has a slope of 2. Hence, the derivative is 2 everywhere. Lastly, what

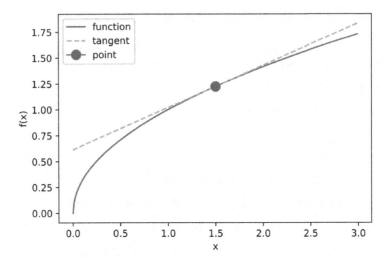

FIGURE 8.9: Function and Tangent at Point

is the derivative of $f(x) = 2x^2$ for some point x? The slope of the tangent to this curve will change as x changes, so this is more complicated than the earlier cases. We will return to this example shortly.

When a function has a derivative at a point x, we say $f(x)$ is *differentiable* at x. A function which is differentiable at each point in its domain is called a *differentiable function*. Since continuity is required for differentiation, this means that a differentiable function over some domain is also a continuous function over this domain. The opposite is not true, however: if a function is continuous over some domain, it does not imply that the function is differentiable over the domain. An example is the absolute value function $f(x) = |x|$, defined as $-x$ when $x < 0$ and x when $x \geq 0$. This function is continuous over its domain $(-\infty, \infty)$, but it is not differentiable at the vertex $x = 0$ where the slope changes sign abruptly. In particular, to the left and right of zero, the slope (and derivative) is -1 and 1, respectively.

The discussion of the derivative thus far has been qualitative. Next, we consider the formal definition, which is entwined with the limit. The derivative of $y = f(x)$ at point x is denoted $y' = f'(x)$ using Lagrange notation, and alternatively as $\frac{dy}{dx}$, $\frac{df(x)}{dx}$, or $\frac{d}{dx}f(x)$ using Leibniz notation. For a constant $h > 0$, the derivative is defined as

$$f'(x) = \lim_{h \to 0} \frac{f(x+h) - f(x)}{h}$$

Breaking down this definition, the fraction on the right side of the equation is the ratio of change in function value, or output, to change in input where h denotes the change. This change h is taken to zero in the limit, and the ratio approaches the tangent at the point x. For this derivative to exist, it needs to be possible to get arbitrarily close to $f(x)$ by moving h arbitrarily close to 0. This means that $f(x)$ must be continuous at x.

Let's apply this definition to compute the derivative of $f(x) = x^2$:

$$\begin{aligned}
f'(x) &= \lim_{h \to 0} \frac{f(x+h) - f(x)}{h} \\
&= \lim_{h \to 0} \frac{(x+h)^2 - x^2}{h} \\
&= \lim_{h \to 0} \frac{x^2 + 2xh + h^2 - x^2}{h} \\
&= \lim_{h \to 0} \frac{2xh + h^2}{h} \\
&= \lim_{h \to 0} \frac{h(2x + h)}{h} \\
&= \lim_{h \to 0} 2x + h \\
&= 2x
\end{aligned}$$

The final step in the calculation applies the limit, sending h to zero. This implies that the derivative $f'(x) = 2x$ is dependent on the point x. For example, $f'(-10) = -20$, $f'(-5) = -10$, $f'(0) = 0$, $f'(5) = 10$, and $f'(10) = 20$. Recalling the graph of the parabola and considering tangents to the points $x = -10, x = -5, x = 0, x = 5$, and $x = 10$, the slope is most negative at $x = -10$, it increases to zero at the parabola vertex $x = 0$ (the tangent is a horizontal line at this point), and it is most positive at $x = 10$.

The derivative of a function can help us find *extrema*, or maxima and minima. In viewing a graph from left to right, suppose the function decreases, flattens, and then increases as with $f(x) = x^2$. The derivative for the corresponding intervals would be negative, zero, and then positive. At the value where the derivative is zero, say $x = c$, $f(c)$ would attain a minimum. Conversely, for a graph that increases, flattens, and then decreases, the derivative would be positive, zero, and then negative. Where the derivative is zero, the function would attain a maximum. This leads to a definition and a useful application of the derivative to find extrema.

A point $x = c$ is a *critical point* of the function $f(x)$ if $f(c)$ exists and either $f'(c) = 0$ or $f'(c)$ does not exist. Function maxima and minima occur at critical points or at the endpoints of intervals. As an example, the function $f(x) = x^2$ has derivative $f'(x) = 2x$ and it is equal to zero when $x = 0$. Hence, $x = 0$ is a critical point. The derivative is negative when $x < 0$ and positive when $x > 0$, indicating that the function reaches a minimum at $x = 0$.

The strategy for finding critical values to locate extrema is very useful, and we will use it later when we select optimal parameter values for regression models. An error function will depend on the parameter values, and we will compute the derivative of the function and look for a critical value. At the critical value, we will find the parameters that minimize the error.

It is often useful to compute derivatives of the derivatives. If f is a differentiable function and f' is its derivative, then the derivative of f', when it exists, is written f'' or f^2 and called the *second derivative* of f. The interpretation of f'' is that it represents the curvature of a function. Similarly, higher-order derivatives can be defined by taking further derivatives as they exist: f''' is the *third derivative* and f^n is the *nth derivative*. For example, we determined the derivative of $f(x) = x^2$ to be $f(x) = 2x$. We could apply the definition of the derivative to compute $f''(x)$, or we can take a shortcut, realizing that this function is a line with slope 2. Thus, $f''(x) = 2$; the second derivative is therefore constant and positive. We can go further, computing the third derivative as $f'''(x) = 0$, since the second derivative is a horizontal line with tangent 0. Next, we will present some rules for computing derivatives. It is faster to use these rules than to apply the definition.

Differentiation Rules

We assume the functions $f(x)$ and $g(x)$ are differentiable on the domain of interest.

Multiplication by a Constant
For constant c and function $f(x)$,

$$\frac{d}{dx}[cf(x)] = cf'(x)$$

In this rule, the constant comes out of the differentiation.

Example:

$$f(x) = 3x$$
$$f'(x) = 3\frac{d}{dx}x$$
$$f'(x) = 3(1) = 3$$

Derivative of a Sum of Functions
For function $h(x) = f(x) + g(x)$, $h(x)$ is differentiable and

$$h'(x) = f'(x) + g'(x)$$

Example: For $f(x) = 3x$ and $g(x) = 4$,

$$h(x) = 3x + 4$$
$$h'(x) = 3\frac{d}{dx}(x) + \frac{d}{dx}(4)$$
$$h'(x) = 3 + 0 = 3$$

Derivative of an Exponential Function

$$\frac{d}{dx}(e^x) = e^x$$

Taking the derivative of the exponential function leaves it unchanged.

Derivative of the Natural Logarithm

$$\frac{d}{dx}\ln(x) = \frac{1}{x}$$

Power Rule

For integer n and function $f(x) = x^n$,

$$f'(x) = nx^{n-1}$$

Example 1:

$$f(x) = x^4$$
$$f'(x) = 4x^3$$

Example 2:

$$f(x) = 3x^5$$
$$f'(x) = 3\frac{d}{dx}(x^5)$$
$$f'(x) = 15x^4$$

Product Rule

For $h(x) = f(x)g(x)$, $h(x)$ is differentiable and

$$h'(x) = f(x)g'(x) + g(x)f'(x)$$

This rule works as first (second)' + second (first)'

Example:

$$h(x) = x^2(3x + 1)$$
$$h'(x) = x^2(3) + (3x + 1)(2x)$$
$$h'(x) = 3x^2 + 6x^2 + 2x$$
$$h'(x) = 9x^2 + 2x$$

Quotient Rule

For $h(x) = \frac{f(x)}{g(x)}$ where $g(x) \neq 0$, $h(x)$ is differentiable and

$$h'(x) = \frac{g(x)f'(x) - f(x)g'(x)}{[g(x)]^2}$$

This rule works as (bottom (top)' – top (bottom)')/(bottom squared)

Example:

$$h(x) = \frac{x}{x^2 + 1}$$

$$h'(x) = \frac{(x^2 + 1) \times (1) - x(2x)}{[x^2 + 1]^2}$$

$$h'(x) = \frac{1 - x^2}{[x^2 + 1]^2}$$

Chain Rule

For a composition of functions $h(x) = f(g(x))$, $h(x)$ is differentiable and

$$h'(x) = f'(g(x))g'(x)$$

This rule works by taking the derivative of the outer function $f(x)$ while keeping the input unchanged, and multiplying by the derivative of the inner function.

Example 1:

$$h(x) = (2 - x)^3$$
$$h'(x) = 3(2 - x)^2(-1) \quad \text{(apply power rule to outer)}$$
$$h'(x) = -3(2 - x)^2$$

For this problem, it is important to realize that $\frac{d}{dx}(2 - x) = -1$.

Example 2:

$$h(x) = e^{2x}$$
$$h'(x) = e^{2x}\frac{d}{dx}(2x)$$
$$h'(x) = 2e^{2x}$$

Example 3:

$$h(x) = ln(1 + e^{2x})$$

$$h'(x) = \frac{1}{1 + e^{2x}} \cdot \frac{d}{dx}(1 + e^{2x})$$

$$h'(x) = \frac{e^{2x}}{1 + e^{2x}} \cdot \frac{d}{dx}(2x)$$

$$h'(x) = \frac{2e^{2x}}{1 + e^{2x}}$$

There are several additional derivative rules for various functions, such as the trigonometric functions. They are out of scope for the purposes of this book.

Partial Derivative

Most real-world problems use functions of several variables. It is valuable to measure function sensitivity to small changes in the inputs. In finance, for example, the risk of a stock will depend on variables like broad stock market performance (e.g., the S&P 500) and interest rates. A risk manager will be interested in the expected change to the stock price when these variables change by some small quantity.

When we consider a function of more than one variable, say $z = f(x, y) = 2x + xy + 3y$, we can take the derivative with respect to each variable while holding the others constant. These derivatives are called *partial derivatives*, and they represent tangent lines in particular directions. The partial derivative notation is slightly different from the single-variable derivative. Let's compute the partial derivatives of z:

$$\frac{\partial z}{\partial x} = 2 + y$$

$$\frac{\partial z}{\partial y} = x + 3$$

Second-order partial derivatives are computed similarly and are denoted

$$\frac{\partial^2 z}{\partial x^2}, \frac{\partial^2 z}{\partial x \partial y}, \frac{\partial^2 z}{\partial y^2}$$

Higher-order derivatives follow analogous notation.

The *gradient* is an important object which holds the partial derivatives of a function at a point. For example, here is the gradient of z:

$$\nabla z(x, y) = \left(\frac{\partial z}{\partial x}(x, y), \frac{\partial z}{\partial y}(x, y) \right)$$

$$\nabla z(x, y) = (2 + y, \quad x + 3)$$

The gradient is a *vector*, which is a quantity with both direction and magnitude. Earlier, we learned that the derivative can be used to find extrema, which can occur where the derivative is zero or undefined (at critical points). For a function of several variables, the gradient can be used in an analogous way to find extrema, as we will see later.

For going deeper on calculus, a masterful pair of books is [26] and [27].

8.4   Probability

Our data will include variables subject to uncertainty or randomness, and this can be treated effectively with tools from probability. A random experiment has an output that cannot be predicted with certainty. When the experiment is repeated a large number of times, however, the average output exhibits predictability. The roll of an unbiased, six-sided die will have an outcome that cannot be predicted, but after a large number of rolls, each number should appear about 17% of the time.

There are three essential ingredients in probability theory:

- The *sample space* is the set of all possible outcomes of the experiment. It is generally denoted Ω. For an experiment where a fair coin is tossed twice, $\Omega = \{hh, ht, th, tt\}$ where h and t are heads and tails, respectively.

- The *events* are outcomes of the experiment. They are subsets of Ω. We can denote the set of all events $\mathcal{A}$. A particular event from this example is $\{hh\}$.

- The probability is a number associated with each event A, written $P(A)$ and falling in $[0, 1]$. The probability of an event increases to 1 as it is more likely. The frequentist idea provides intuition: for a large number of repeated trials, the probability of an event will be the number of times the event occurs divided by the number of trials.

A *random variable* is a quantity subject to random events, and it depends on the outcome of the experiment. Mathematically, it is a mapping from Ω to a measurable space (usually $\mathbb{R}$). Here are some examples:

- Let X be the number of heads tossed after two flips of a fair coin

- Let Z denote the number of points scored by the winning team in a Super Bowl

Discrete and Continuous Random Variables

A random variable is either a *discrete random variable* or a *continuous random variable*, and this is determined by the number of possible values it can take. Note that the principle is the same as discrete and continuous data that we encountered earlier: a discrete random variable takes a finite or countable number of values, while a continuous random variable takes an uncountable number of values. A random variable representing the sum from rolling two dice is discrete, and so is the cumulative number of visitors to `www.google.com` at any given time. Continuous random variables depend on quantities taking real-valued measurements like time, distance, and volume. An example would be the distance of citizens from a police station. Since most problems are now solved with computers which have finite memory, values are rounded and this results in discrete measurement. However, treating variables with high cardinality as continuous can be convenient and very effective.

The simplest probabilities to compute involve discrete random variables with equally likely outcomes. When rolling a fair, six-sided die, there are six, equally likely outcomes. The probability of rolling a five, for example, is equal to the number of possible ways to arrive at success divided by the number of possible outcomes: $P(5) = 1/6$. When rolling two such dice, the probability of the sum Y being equal to five follows the same approach:

$$P(Y = 5) = \frac{\#(\text{ways of achieving 5})}{\#(\text{possible outcomes})}$$
$$= \frac{|\{(4,1),(1,4),(2,3),(3,2)\}|}{36}$$
$$= \frac{4}{36} = \frac{1}{9}$$

Conditional Probability

When computing a probability, it is important to consider what prior information is available. From the dice-rolling example, we used the fact that each side of a die would appear with equal probability. This allowed us to say that each roll produces six equiprobable outcomes $\{1, 2, \ldots, 6\}$, and two rolls produce $6 \times 6 = 36$ equiprobable outcomes. Now let's ask the same question with some additional data. What is the probability that the sum of two dice is 5 given that one of the dice is showing 1? To do this, we can return to our earlier formula of $\#(\text{ways of achieving success})/\#(\text{possible outcomes})$, but we need to update these expressions. In the numerator, it is no longer possible to roll (2,3) or (3,2). In the denominator, there are no longer 36 possible outcomes, but rather 12 outcomes involving a 1: $\{((1,1),(1,2),\ldots,(1,6),(6,1),(5,1),\ldots,(1,1))\}$

$$P(Y = 5) = \frac{\#(\text{ways of achieving 5 given a 1})}{\#(\text{possible outcomes})}$$

$$= \frac{|\{(4,1),(1,4)\}|}{12}$$

$$= \frac{2}{12} = \frac{1}{6}$$

When we compute the probability given available information, we call this the *conditional probability*. Mathematically, this is written $P(Y|X)$ and read "the probability of Y given X." In the case above, $Y = 5$ represented two dice summing to 5, and $X = 1$ represented one of the dice having value 1. More generally, the formula is

$$P(Y|X) = \frac{P(X \cap Y)}{P(X)}$$

where $P(X \cap Y)$ denotes the probability of both X and Y occurring. For our example, the numerator represents the probability that the two rolls sum to 5, with the restriction that at least one die is 1. We saw there are two such outcomes out of 36 possible outcomes, for a probability of 2/36. The denominator works out to 12 possible rolls involving one or more 1s out of 36 possible outcomes, for a probability of 12/36. The calculation is done like this:

$$P(Y = 5|X = 1) = \frac{(2/36)}{(12/36)}$$

$$= \frac{2}{12} = \frac{1}{6}$$

When the outcome of X does not influence the outcome of Y, we say that X and Y are *independent*. This can be written $P(Y|X) = P(Y)$. In data science, the goal is to find random variables $X_1, X_2, \ldots, X_p$ which predict Y. Predictors which are independent of Y would not be helpful. We will see later that prediction relies heavily on conditional probability, as important data is included to produce refined results.

Probability Distributions: Discrete Case

We are sometimes in the fortunate position of being able to state the probability of the random variable taking each possible value. This increases the certainty of the outcome. The *probability mass function* (pmf) for a discrete random variable is a function that takes each possible value as input and returns its probability as an output. For a random variable Y equal to the value showing on a fair, six-sided die, the pmf looks like this:

$$f(y) = P(Y = y) = \begin{cases} 1/6, & \text{if } y = 1 \\ 1/6, & \text{if } y = 2 \\ 1/6, & \text{if } y = 3 \\ 1/6, & \text{if } y = 4 \\ 1/6, & \text{if } y = 5 \\ 1/6, & \text{if } y = 6 \end{cases}$$

Y denotes the random variable, while y denotes the specific value taken. Given equal probabilities for each of the outcomes, this distribution is an example of a *discrete uniform distribution*.

The *cumulative distribution function* (cdf) for a discrete random variable is a function that takes the possible values as input and returns the probability of realizing that value or lower as output. The cdf for the die looks like this:

$$F(y) = P(Y \le y) = \begin{cases} 1/6, & \text{if } y = 1 \\ 2/6, & \text{if } y = 2 \\ 3/6, & \text{if } y = 3 \\ 4/6, & \text{if } y = 4 \\ 5/6, & \text{if } y = 5 \\ 1, & \text{if } y = 6 \end{cases}$$

As an example, the probability of rolling 2 or lower is $P(Y \le 2) = P(Y = 1) + P(Y = 2) = 2/6$. For a quick understanding of the random variable, it is often useful to graph the pmf and cdf with the input on the x-axis and the output on the y-axis.

To connect these concepts to our work with data, we take the view that various data columns are random variables. The observations are particular realizations of these random variables. For example, we might collect realizations of gender for a sample of data. This is a discrete random variable, and the realized percentages of each gender constitute the pmf.

Probability Distributions: Continuous Case

For a continuous random variable, there is zero probability that it will achieve any particular value. For some intuition, consider the case where there are n outcomes which have equal probability $1/n$. When $n = 6$, each outcome has probability $1/6$. After taking the limit $n \to \infty$, the probability of each outcome is $1/\infty = 0$. Rather than measuring the probability of particular values, we measure the probability that a continuous random variable assumes a value in a range, say $P(0 < Y < 1)$.

Since a continuous random variable takes infinitely many values, a probability mass function is not appropriate. Instead, a *probability density function* (pdf) is used. The fundamental difference between the pmf and the pdf is that the pdf value at a single point does not have meaning. For computing the cumulative probability $F(y) = P(Y \leq x)$, the sum will not make sense; instead, the area under the pdf curve is computed by using *integration*. We won't discuss integrals here, but rather will review some continuous random variables where integrals are not explicitly needed. First, we start with some examples of discrete random variables.

Bernoulli Distribution

The *Bernoulli distribution* is a discrete probability distribution for a random variable that takes value 1 (this is a success) with probability p and value 0 with probability $1 - p$ (this is a failure). As there are two possible outcomes, it can model the result of a coin flip, for example. For random variable Y following a Bernoulli distribution, the pmf is

$$f(y) = \begin{cases} p, & \text{if } y = 1 \\ 1 - p, & \text{if } y = 0 \end{cases}$$

Binomial Distribution

The *binomial distribution* is a discrete probability distribution that counts the number of successes out of n trials following a Bernoulli distribution with success probability p. The distribution can be written compactly as $Y \sim Bin(n, p)$. As an example, we can compute the probability of flipping one head out of ten coin flips, where the probability of heads is 0.75. This can be calculated like this:

$f(1) = \#(\text{ways to flip 1 head}) \times (\text{prob of 1 success}) \times (\text{prob of 9 failures})$

The pmf expresses this more generally for y successes:

$$f(y) = \binom{n}{y} p^y (1 - p)^{n-y} \;\; for \;\; y = 0, 1, \ldots, n$$

where $\binom{n}{y} = \frac{n!}{y!(n-y)!}$ is the number of ways that y can happen. For example, if $n = 10$ and $y = 1$, then a success can occur in any trial and $\binom{10}{1} = 10$.

Uniform Distribution

The *uniform distribution* is a continuous probability distribution on the interval (a, b). It is denoted $U(a, b)$ and has pdf:

$$f(y) = \begin{cases} \frac{1}{b-a}, & \text{if } a < y < b \\ 0, & \text{if } otherwise \end{cases}$$

We can envision the pdf as a rectangle having length $b - a$, height $1/(b - a)$, and area 1. The cdf $F(y)$ is found by computing the area under the rectangle starting at $Y = a$ and ending at $Y = x$:

$$F(y) = \begin{cases} 0, & \text{if } y \leq a \\ \frac{y-a}{b-a}, & \text{if } a < y < b \\ 1, & \text{if } y \geq b \end{cases}$$

Normal Distribution

The *normal distribution* or *Gaussian distribution* is a continuous probability distribution on the interval $(-\infty, \infty)$. It is a famous distribution with a bell-shaped pdf. Many random variables follow a normal distribution or something approximately normal. Additionally, the sum of a large number of independent, identically distributed (i.i.d.) random variables follow an approximately normal distribution. The normal distribution is fully defined by two parameters: the *mean* μ and the *variance* σ^2. It is commonly denoted $\mathcal{N}(\mu, \sigma^2)$ for real numbers μ and σ^2 with $\sigma > 0$. The mean measures the center of the data while the variance measures the spread of the data. We will discuss these quantities in more detail later. The pdf of the normal distribution is:

$$f(x) = \frac{1}{\sigma\sqrt{2\pi}} \exp\left(-\frac{(x - \mu)^2}{2\sigma^2}\right), \quad -\infty < x < \infty$$

The cdf of the normal distribution is computed numerically, as there is no closed-form solution. In practice, values are easily obtained through software.

Standard Normal Distribution

The *standard normal distribution* is a continuous probability distribution that derives from the normal distribution. A normally distributed random variable X can be converted to a random variable Z with standard normal distribution through the formula:

$$Z = \frac{X - \mu}{\sigma}$$

This process is called *standardization*. The value Z is called a *z-score* and it represents the distance of X from the mean in units of standard deviation. Standardization allows for comparing variables which are normally distributed with different parameters. The mean and variance of the standard normal are zero and one, respectively, and the distribution is denoted $\mathcal{N}(0, 1)$. The pdf of the standard normal distribution is:

$$f(z) = \frac{1}{\sqrt{2\pi}}\, e^{-z^2/2}, \quad -\infty < z < \infty$$

The cdf is computed numerically. The following code snippet shows some probability calculations for a standard normal. It uses the `scipy` module which includes functionality for statistics, optimization, linear algebra, and more.

A Note on numpy

The numpy module also supports work with random variables; it is an essential package for numerical computing in Python. Numpy defines a highly efficient object for arrays called the *NumPy array*. We will work with NumPy arrays throughout the book.

```
from scipy import stats

# compute the cdf when z=0
print('P(Z < 0):', stats.norm.cdf(0))

# compute as 1 - F(0)
print('P(Z > 0):', 1-stats.norm.cdf(0))

# compute F(1) and subtract left tail F(-1)
print('P(-1 < Z < 1):',
round(stats.norm.cdf(1) - stats.norm.cdf(-1), 3))

print('P(-2 < Z < 2):',
round(stats.norm.cdf(2) - stats.norm.cdf(-2), 3))

print('P(-3 < Z < 3):',
round(stats.norm.cdf(3) - stats.norm.cdf(-3), 3))

OUTPUT:

P(Z < 0): 0.5
P(Z > 0): 0.5
P(-1 < Z < 1): 0.683
P(-2 < Z < 2): 0.954
P(-3 < Z < 3): 0.997
```

The output indicates some interesting facts about the standard normal distribution. First, half of the probability is below $Z=0$ and half is above, by symmetry. Roughly 68% of the probability is between –1 and 1, or one standard deviation of the mean (which is zero). There is about 95% of the probability within two standard deviations of the mean, and 99.7% (nearly all) of the probability within three standard deviations of the mean. These numbers are so common in probability and statistics that they are worth remembering.

For simulation and other applications, it is valuable to generate draws of random variables from various distributions. This is easily accomplished with software, which uses a *seed* paired with a pseudorandom number generator. For a given seed, it will generate a fixed sequence of numbers. Here is an example of generating standard normals, where two values are taken (`size=2`) at a time in two separate runs. The values are completely different on each run.

```
print('run 1:', stats.norm.rvs(size=2))
print('run 2:', stats.norm.rvs(size=2))
```

```
run 1: [ 0.85228509 -0.70707096]
run 2: [-0.93165719  0.88666088]
```

It is often desirable to regenerate the same values on repeated runs. This example gives a demonstration:

```
import numpy as np
```

```
np.random.seed(seed=314)
print('run 1:', stats.norm.rvs(size=2))
np.random.seed(seed=314)
print('run 2:', stats.norm.rvs(size=2))
```

```
run 1: [0.16608544 0.78196448]
run 2: [0.16608544 0.78196448]
```

Figure 8.10 shows a histogram based on 100,000 random draws from the standard normal distribution. Notice the bell shape and domain where most values fall (between –3 and 3). The source code is minimal.

```
import seaborn as sns
```

```
sns.histplot(stats.norm.rvs(size=100000))
```

Central Limit Theorem

For a large number of i.i.d. random variables $X_1, X_2, \ldots, X_n$ with mean μ and variance σ^2, their average $\bar{X}_n$ converges to a normal distribution. As the name suggests, this is an extremely important result, as averaging a sufficient number of i.i.d. random variables with any distribution will produce something normally distributed. In particular, this is very useful when collecting samples of data which are often drawn from an unknown distribution. Taking things one step further, if we standardize $\bar{X}_n$ by subtracting its mean and dividing by

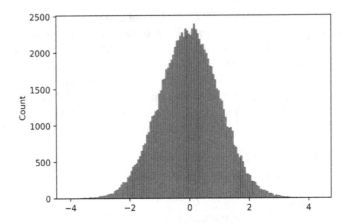

FIGURE 8.10: Histogram of Standard Normal Draws

its standard deviation $\sigma/\sqrt{n}$, the resulting variable will converge to a standard normal. Let's state the *central limit theorem*:

As $n \to \infty$,

$$\sqrt{n}\left(\frac{\bar{X}_n - \mu}{\sigma}\right) \to \mathcal{N}(0,1)$$

Probability is a vast subject and we have only scratched the surface. See [28] for a broader treatment of the subject.

8.5   Matrix Algebra

Matrix, Vector, and Scalar

A *matrix* $\mathbf{A}$ is a rectangular array of numbers with m rows and n columns written

$$\mathbf{X} = \begin{bmatrix} a_{11} & a_{12} & a_{13} & \cdots & x_{1n} \\ a_{21} & a_{22} & a_{23} & \cdots & x_{2n} \\ \vdots & \vdots & \vdots & \ddots & \vdots \\ a_{m1} & a_{m2} & a_{m3} & \cdots & a_{mn} \end{bmatrix}$$

Matrices (plural of matrix) will be useful when we organize our data for calculations. They also provide a compact notation for solving systems of equations, which we encounter in the upcoming regression modeling.

The number of rows m and columns n of a matrix comprise the dimensions of the matrix: $m \times n$. The dimensions are an extremely important attribute of a matrix, as they define which operations are possible.

Here is a 2×3 matrix:

$$\mathbf{X} = \begin{bmatrix} 1 & 5 & 0 \\ 2 & 12 & -1 \end{bmatrix}$$

A matrix with one row is called a *row vector*, while a matrix with one column is called a *column vector*. A vector with only one element is a *scalar*.

Here is a row vector of length 3 and a column vector of length 4, respectively:

$$\mathbf{r}^{\mathbf{T}} = \begin{bmatrix} 1 & 5 & 0 \end{bmatrix} \qquad \mathbf{c} = \begin{bmatrix} 2 \\ 7 \\ 3 \\ -1 \end{bmatrix}$$

Matrix Addition and Subtraction

Two matrices may be added or subtracted if their dimensions match. When adding or subtracting two matrices $\mathbf{A}$ and $\mathbf{B}$, their corresponding elements are added or subtracted in an element-wise fashion like this:

$$\mathbf{A} = \begin{bmatrix} 1 & 5 & 0 \\ 2 & 12 & -1 \\ 4 & 12 & 3 \end{bmatrix} \qquad \mathbf{B} = \begin{bmatrix} 1 & 4 & 0 \\ -1 & 0 & 6 \\ 0 & 1 & 0 \end{bmatrix}$$

$$\mathbf{A} + \mathbf{B} = \begin{bmatrix} 1 & 5 & 0 \\ 2 & 12 & -1 \\ 4 & 12 & 3 \end{bmatrix} + \begin{bmatrix} 1 & 4 & 0 \\ -1 & 0 & 6 \\ 0 & 1 & 0 \end{bmatrix}$$

$$= \begin{bmatrix} 1+1 & 5+4 & 0+0 \\ 2+(-1) & 12+0 & -1+6 \\ 4+0 & 12+1 & 3+0 \end{bmatrix}$$

$$= \begin{bmatrix} 2 & 9 & 0 \\ 1 & 12 & 5 \\ 4 & 13 & 3 \end{bmatrix}$$

Matrix Multiplication

For $m \times n$ matrix $\mathbf{E}$ and $n \times p$ matrix $\mathbf{F}$, two products $\mathbf{EF}$ and $\mathbf{FE}$ may be formed, and they are generally different. The dimensions of $\mathbf{E}$ and $\mathbf{F}$ determine if the product is valid, and if so, the dimensions of the product. A short mnemonic can be used: if $\mathbf{X}$ is $a \times b$ and $\mathbf{Y}$ is $c \times d$, then for $\mathbf{XY}$ we write: $(a \times b) \times (c \times d)$. The inner dimensions are b and c, while the outer dimensions are a and d. The inner dimensions must match for the product to be valid. The product will have the outer dimensions.

For the case of $\mathbf{E}$ and $\mathbf{F}$, the dimension calculation looks like $(m \times n) \times (n \times p)$. The inner dimensions are both n, and so the product is valid. The outer dimensions are m and p, and so $\mathbf{EF}$ will be $m \times p$. This means that the product will have the same number of rows as the left matrix and the same number of columns as the right matrix.

Returning to our numeric example, $\mathbf{C} = \mathbf{AB}$ can be formed as follows: For each element of $\mathbf{C}$, written c_{ij}, select row i from $\mathbf{A}$ and column j from $\mathbf{B}$. Next compute the *inner product* of these two vectors as follows:

$$c_{ij} = \sum_{k=1}^{p} a_{ik} b_{kj}$$

For this example,

$$
\begin{aligned}
c_{11} &= \sum_{k=1}^{p} a_{1k} b_{k1} \\
&= (a_{11} \times b_{11}) + (a_{12} \times b_{21}) + (a_{13} \times b_{31}) \\
&= (1 \times 1) + (5 \times (-1)) + (0 \times 0) \\
&= -4
\end{aligned}
$$

Computing the inner products for the remaining elements produces

$$
\mathbf{AB} = \begin{bmatrix} 1 & 5 & 0 \\ 2 & 12 & -1 \\ 4 & 12 & 3 \end{bmatrix} \begin{bmatrix} 1 & 4 & 0 \\ -1 & 0 & 6 \\ 0 & 1 & 0 \end{bmatrix} = \begin{bmatrix} -4 & 4 & 30 \\ -10 & 7 & 72 \\ -8 & 19 & 72 \end{bmatrix}
$$

Next, let's look at a code example. We create the matrices $\mathbf{A}$ and $\mathbf{B}$ as NumPy arrays. The operator @ supports matrix multiplication. Note that the operator $*$ is element-wise multiplication and not matrix multiplication. The character \n inserts carriage returns for nicer output.

```
A = np.array([[1,5,0],[2,12,-1],[4,12,3]])
B = np.array([[1,4,0],[-1,0,6],[0,1,0]])

print('A: \n',A)
print('')
```

```
print('B: \n',B)
print('')
print('AB: \n', A@B)
```

OUTPUT:

```
A:
 [[ 1  5  0]
  [ 2 12 -1]
  [ 4 12  3]]

B:
 [[ 1  4  0]
  [-1  0  6]
  [ 0  1  0]]

AB:
 [[ -4   4  30]
  [-10   7  72]
  [ -8  19  72]]
```

Let's define another matrix **D** with dimensions that do not support matrix multiplication with **A**. Then we can see how Python handles such a case.

```
D = np.array([[1,4,0],[-1,0,6]])
```

```
print('D: \n', D)
print('')
print('A dimensions:', A.shape)
print('D dimensions:', D.shape)
print('AD: \n', A@D)
```

OUTPUT:
```
D:
 [[ 1  4  0]
  [-1  0  6]]

A dimensions: (3, 3)
D dimensions: (2, 3)

ValueError: matmul: Input operand 1 has a mismatch
in its core dimension 0, with gufunc signature
(n?,k),(k,m?)->(n?,m?) (size 2 is different from 3)
```

The shape attribute retrieves the dimensions of a matrix. Since **A** has 3 columns while **D** has 2 rows, the product **AD** will be invalid. Python throws a

`ValueError` indicating a dimension mismatch when calling the `matmul` method
(the `@` operator is shorthand for this method). Operations with invalid dimensions are a common source of bugs involving matrix algebra. Checking
the dimensions of each matrix is highly recommended.

Transpose

The transpose of matrix $\mathbf{A}$ is a matrix denoted $\mathbf{A}^{\mathbf{T}}$ or $\mathbf{A}'$. $\mathbf{A}^{\mathbf{T}}$ is obtained by
taking the rows (or columns) of $\mathbf{A}$ to be the columns (or rows) of $\mathbf{A}^{\mathbf{T}}$. Here
is an example:

$$\mathbf{A} = \begin{bmatrix} 1 & 5 & 0 \\ 2 & 12 & -1 \\ 4 & 12 & 3 \end{bmatrix}$$

$$\mathbf{A}^{\mathbf{T}} = \begin{bmatrix} 1 & 2 & 4 \\ 5 & 12 & 12 \\ 0 & -1 & 3 \end{bmatrix}$$

For example, the first row of $\mathbf{A}$, $[1, 5, 0]$, became the first column of $\mathbf{A}^{\mathbf{T}}$. Note
that $[1, 5, 0]$ is a row vector, and taking its transpose made it a column vector.
Taking the transpose of a column vector makes it a row vector.

When a matrix $\mathbf{M}$ is equal to its transpose $\mathbf{M}^{\mathbf{T}}$, it is said to be *symmetric*.
An example is

$$\mathbf{M} = \begin{bmatrix} 1 & 2 & 4 \\ 2 & 12 & -6 \\ 4 & -6 & 3 \end{bmatrix}$$

Symmetric matrices can reduce the amount of storage and computation
needed, due to repeated values in predictable locations (here, $m_{12} = m_{21}$,
$m_{13} = m_{31}$, and $m_{23} = m_{32}$).

Matrix Diagonal

The diagonal of an $n \times n$ matrix $\mathbf{M}$ includes the elements $m_{11}, m_{22}, \ldots, m_{nn}$.
That is, they begin at the top left and proceed down to the bottom right.
From the above example, the elements on the diagonal of $\mathbf{M}$ are 1, 12, and 3.

Identity

The *identity matrix* $\mathbf{I_n}$ is a square $n \times n$ matrix with 1s on the diagonal
and 0s elsewhere. The subscript n is sometimes suppressed. Multiplying an
appropriately sized matrix $\mathbf{M}$ by $\mathbf{I_n}$ (on the left or right) will reproduce $\mathbf{M}$.

This serves the same role as the number 1 in multiplication. Here is a 3×3 example:

$$\mathbf{I_3} = \begin{bmatrix} 1 & 0 & 0 \\ 0 & 1 & 0 \\ 0 & 0 & 1 \end{bmatrix}$$

Zero Vector and Matrix

The *zero vector* and *zero matrix* contain all zeros. Here are examples of each:

$$\mathbf{0^T} = \begin{bmatrix} 0 & 0 & 0 \end{bmatrix}$$

$$\mathbf{0} = \begin{bmatrix} 0 & 0 & 0 \\ 0 & 0 & 0 \\ 0 & 0 & 0 \end{bmatrix}$$

Inverse

The inverse $\mathbf{M}^{-1}$ of a square matrix $\mathbf{M}$ is a matrix such that $\mathbf{MM}^{-1} = \mathbf{M}^{-1}\mathbf{M} = \mathbf{I}$. The inverse is unique, and it serves the same role as in arithmetic. Finding the inverse of a matrix is done efficiently by computer (e.g., Python has a function `numpy.linalg.inv()`), as it involves solving a system of linear equations. *Gaussian elimination* is one popular method for inverting a matrix. A matrix with an inverse is called an *invertible matrix*; it is otherwise called a *non-invertible matrix*. For further details, an excellent text is [29].

8.6 Chapter Summary

We started this chapter with an introduction to set theory. Sets, or collections of objects, are everywhere in data science, from sets of product users to sets of recommendations. It is often important to compare and combine sets; operations including the complement, union, and intersection were discussed. Sets are also essential building blocks of functions and probability.

A function is a mapping that takes an element of a set and assigns an element of a second set as output. In data science, we use functions to transform input data to (hopefully) more useful output data. We might also perform a sequence of transformations, which is an example of function composition. The inverse function of a function serves to undo the function. We discussed checking if a curve is a function (each input must map to only one output), and if a function has an inverse (it must be one-to-one and onto).

Next, we investigated the monotonicity and convexity properties of a function. A monotonic function is either entirely nonincreasing or nondecreasing. A convex function forms a cup shape. Convexity simplifies the task of searching for the global minimum value on a domain.

The final topic on functions was a survey of different function types. Polynomials include linear functions, which feature prominently in regression analysis. The elementary transcendental functions include trigonometric functions, the exponential function, and the logarithm.

Many functions used in machine learning are differentiable functions. To build conceptual understanding, we briefly reviewed the important concepts of limit, continuity, and derivative from differential calculus. The limit is the notion of getting arbitrarily close to some value by changing the input value. A function is continuous at point $f(\xi)$ if it is possible to get arbitrarily close by moving x close to some point ξ. A function is continuous on an interval if it is continuous at each point in the interval.

The derivative measures the change in function output with respect to a change in input. It is equivalent to the tangent to the function at a given point after taking a limit. A function which is differentiable at each point in its domain is called a differentiable function.

The derivative has several useful applications and it can help us understand the underlying function. Differentiability implies continuity, but the reverse is not true. The derivative can help us find extrema through use of critical points; these are points where the derivative is zero or undefined. Differentiation rules are faster to apply than the definition of the derivative; we reviewed some of these rules. The partial derivative measures specific tangents to a function of multiple variables.

As our data is subject to randomness, we briefly reviewed probability theory, which rests on a thorough mathematical framework. The essential ingredients are the triplet of sample space, events, and probability. A random variable is a quantity subject to random events. It is a mapping from the sample space to a measurable space. A random variable may be discrete (finite or countable number of outcomes) or continuous (uncountable number of outcomes).

The probability that an event will occur is the number of ways of achieving success divided by the number of possible outcomes. When information is provided, this can be included in probability calculations; the probability given available information is called the conditional probability. When one variable does not affect the probability of another, they are independent.

We reviewed examples of discrete random variables (discrete uniform, Bernoulli, and Binomial) and continuous random variables (uniform, Normal, and Standard normal). Discrete random variables use a probability mass function to specify the probability at each possible value. Continuous random variables use a probability density function to specify the probability over a range of values, as the probability of any particular value is zero. The cumulative distribution function is a function that takes the possible values as input and returns the probability of realizing that value or lower as output.

Many variables follow a normal distribution or something approximately normal. The sum of a large number of independent, identically distributed random variables follows an approximately normal distribution by the central limit theorem. Is it often useful to convert a normally distributed random variable to a standard normal using standardization. Computing probabilities of random variables and simulating random variables can be easily accomplished through software.

A matrix is a rectangular array of numbers. The number of rows and columns of a matrix, or dimensions, will be important when performing operations. We reviewed matrix addition, subtraction, and multiplication. Matrix multiplication is not commutative (**AB** does not always equal **BA**). Addition and subtraction are done element-wise, while multiplication is done with inner products. The transpose of a matrix is formed by interchanging rows for columns (or vice versa). The identity of a matrix is a square matrix with 1s on the diagonal and 0s elsewhere. The inverse of a square matrix **M** will produce the identity matrix when it is multiplied by **M** on the left or right. When a matrix has an inverse, it is called invertible. The inverse of a matrix is best implemented with software.

This chapter provides the necessary mathematical preparation for the remainder of the book. In the next chapter, we will briefly review the essential statistical concepts needed. We will see that statistics uses probability as a foundation.

8.7 Exercises

Exercises with solutions are marked with Ⓢ. Solutions can be found in the `book_materials` folder of the course repo at this location:

https://github.com/PredictioNN/intro_data_science_course/

1. Consider two sets A and B. How could you prove they are equal?

   Consider the sets U, X, Y, and Z below, where U is the universal set.

   $U = \{a, b, c, d, e, f, g\}$
   $X = \{a, b, c\}$
   $Y = \{d, e, f\}$
   $Z = \{b, c, g\}$

2. Ⓢ What is the union of X and Y?

3. Ⓢ What is the intersection of Y and Z?

4. Ⓢ Does the function $f(x) = 7x^2$ have an inverse? Explain your reasoning.

5. Find the inverse of the function $f(x) = \sqrt{2x}$. Graph each function and notice the symmetry about the line $y = x$.

6. Ⓢ Explain the limit of a function.

7. Ⓢ What is the limit of the function $f(x) = \frac{1}{2x-1}$ as x approaches $1/2$?

8. What is the limit of the function $f(x) = \frac{sin(x)}{3x+1}$ as x approaches 0?

9. Write a function that takes n as input and returns the sum $S_n = 1 + \frac{1}{1!} + \frac{1}{2!} + \ldots + \frac{1}{n!}$ for $n > 2$. Call the function with $n = 1000$ and compare the result to e.

10. Consider the stepwise function f defined below. Indicate the domain where f is continuous.

$$f(x) = \begin{cases} 0 & x \le 0 \\ x & x > 0 \end{cases}$$

11. Ⓢ Explain the derivative of a function.

12. Consider the stepwise function f defined below. Indicate the domain where f is differentiable.

$$f(x) = \begin{cases} -x & x \le -1 \\ x & x > -1 \end{cases}$$

13. True or False: If a continuously differentiable function has a positive derivative on a domain, then it is monotonically increasing on that domain.

14. True or False: If a function is continuous on $[a, b]$ then it must also be differentiable on $[a, b]$.

15. Calculate the derivative of $f(x) = 5$

16. Ⓢ Calculate the derivative of $f(x) = 6x^2 + 1$

17. Calculate the derivative of $f(x) = e^{x^2+1}$

18. Ⓢ Calculate the derivative of $f(x) = 2x^3 e^{2x}$

19. Calculate the derivative of $f(x) = log(4x^2)$

20. Ⓢ Apply the derivative to find the maximum of the function $f(x) = -4x^2 + 5x$ on the domain $(-\infty, \infty)$. For which value of x does the maximum occur?

21. Ⓢ Compute the gradient of the function $f(x, y) = 2x + 3y^2 + 1$

22. ⓈA couple has two children. One child is known to be a girl. What is the probability that the other child is a girl?

23. A fair coin is flipped five times. What is the probability of three heads?

24. ⓈA random variable U follows a Uniform(0,1). What is the probability that U is between 1/3 and 2/3?

25. ⓈTrue or False: For a random variable X following a normal distribution, the probability that X=0 is zero.

26. ⓈA random variable X follows a normal distribution with mean 10 and standard deviation 2. What is the probability that X is between 8 and 14?

27. ⓈA random variable follows a Weibull distribution with shape parameter 1 and scale parameter 2. One thousand independent draws are made and the mean value is computed. This is repeated 100,000 times. What is the approximate distribution of the means?

28. Consider symmetric matrix A and its transpose B. True or False: A is equal to B.

29. True or False: It is possible to add two matrices if they have different dimensions.

30. ⓈConsider a vector $\mathbf{v}$ which contains zeroes in the first 99 positions and a value of 1 in the final position. A second vector $\mathbf{w}$ contains a 1 in the first position and zeroes in the next 99 positions. What is the inner product of $\mathbf{v}$ and $\mathbf{w}$?

31. Verify that for a square $n \times n$ matrix $\mathbf{M}$ and the identity matrix $\mathbf{I_n}$, the matrix product $\mathbf{MI_n} = \mathbf{I_nM} = \mathbf{M}$.

32. Use numpy to create 3×3 matrices $\mathbf{A}$ and $\mathbf{B}$ populated with random values. Calculate $\mathbf{A} + \mathbf{B}$ and $\mathbf{A} - \mathbf{B}$. Verify the results by hand.

33. Use numpy to create a 2×3 matrix $\mathbf{A}$ and a 3×3 matrix $\mathbf{B}$. Populate $\mathbf{A}$ and $\mathbf{B}$ with random values. Calculate $\mathbf{AB}$. Verify the results by hand.

9

Statistics Preliminaries

To save on resources (time, money, etc.) we draw samples of data from a population of size N. A *statistic* is a quantity computed from a data sample. The hope is that the statistic is representative of the population. As data science is grounded in the collection and analysis of data samples, we have a great need to work with statistics. This section will provide a summary of *descriptive statistics* and *inferential statistics*. Descriptive statistics are useful for understanding the properties of a data sample. Inferential statistics allow for drawing conclusions about a population based on a sample. A common question in inference, for example, is whether a predictor's coefficient is significantly different from zero; if so, this indicates that the predictor is important.

9.1 Descriptive Statistics

Our data samples will be drawn from random variables which may be discrete or continuous. Each variable may take a single value at one extreme to an uncountably large number of values at the other extreme. As the sample size grows, it becomes impossible to look at all of the data and understand its distribution. Computing statistics will help in summarizing this data, and we outline several different statistics in what follows.

Frequency Count

For data with low cardinality (say below 50 unique levels), we can count the number of times that each value is taken. These occurrences can be called *frequency counts* or simply *counts*. It is also useful to calculate the percentage of times each value is taken. The values with their percentages are then equivalent to the probability mass function. For example, given observed data $\mathcal{S} = (1, 1, 1, 2, 3, 5000)$, the values and associated counts are $F = \{1 : 3, 2 : 1, 3 : 1, 5000 : 1\}$. The values, counts, and percentages can be assembled in a table like the following:

value	count	percent(%)
1	3	50.0
2	1	16.7
3	1	16.7
5000	1	16.7
Total	6	100

For data with high cardinality, the values can be placed into bins, and then counts and percentages can be computed for each bin. This is what is done when creating a histogram. The table below summarizes counts and percentages for a large number of values measured to the thousandths place, such as 0.412 and 15.749. The bin $[0, 5)$ will contain values including the left endpoint of 0 and excluding the right endpoint of 5.

bin	count	percent(%)
[0,5)	525	40
[5,10)	173	25
[10,15)	298	15
[15,20)	97	18
[20,25)	45	2
Total	1138	100

For a large number of bins, the data distribution can be better understood with a graph of the bins on the x-axis and their occurrences on the y-axis (as in a vertical bar chart).

Central Tendency

The *central tendency* or *center* of the data is one of the most common and useful attributes. It can be measured by the *mode*, *median*, or *mean*. The mode is the most common value taken by the variable. Given frequency counts as in F, the mode is equal to the key with highest value (1 in this case). The mode will not be influenced by outliers, as they are rare by definition. A shortcoming of the mode is that it is purely frequency based.

The median is the value in the "middle" of the data. To be more precise, we can first define the pth percentile of the data to be the value which exceeds p percent of the data. Then the median can be defined as the 50th percentile of the data; half of the values are above and below the median. For an odd number of data points, this can be found by sorting the values and taking the middle point. For an even number of data points, the median is computed as the midpoint of the two middle values. For the example S, the median will be 1.5, which is the midpoint of 1 and 2. Like the mode, the median is not influenced by outliers. The median gives a better estimate of center since it uses the ranking of values.

The average or *sample mean* of the data applies equal weight to each observation in estimating where the data "balances." It is a statistic that estimates the *population mean*, denoted μ. The sample mean is computed by summing the values and dividing by the number of observations n. For values $x_1, x_2, \ldots, x_n$ drawn from random variable X, the formula is

$$\bar{x} = \frac{1}{n} \sum_{i=1}^{n} x_i$$

Extreme observations will have a large influence on the sample mean; for measuring the center in such a case, the median will be better. From our example $\mathcal{S}$, the sample mean is approximately 835 due to the large outlier. While this value might give a sense of where the data balances, it does not represent most of the data. When we see the median and sample mean together for a dataset, their similarity or difference provides additional information. When the two statistics are similar, this indicates symmetry in the data distribution. When the sample mean is much larger (smaller) than the median, this indicates one or more large (small) outliers.

Spread

The *dispersion* or *spread* of the data gives a sense of how the values are distributed about a central value. For a spread of zero, all values would be the same; such a variable would not be random at all. The spread is a measure of the variable's uncertainty, and several statistics are used: the *range*, the *interquartile range* (IQR), the *variance*, and the *standard deviation*.

The range is the difference between the maximum and minimum values of the data. It is simple to calculate and intuitive, but influenced by outliers, as we see with our sample data where $range(\mathcal{S}) = 5000 - 1 = 4999$.

For measuring the interquartile range, we first sort the data from lowest to highest and divide the sorted values into four equal parts, or *quartiles*. The lowest quartile of the data ranges from the minimum to $Q1$ (the 25th percentile), the second quartile ranges from $Q1$ to $Q2$ (the 50th percentile or median), the third quartile ranges from $Q2$ to $Q3$ (the 75th percentile), and the fourth quartile ranges from $Q3$ to $Q4$ (the maximum). The middle 50% of the data is then found between $Q1$ and $Q3$.

The interquartile range is calculated as $Q3 - Q1$, which represents the range covered by the middle 50% of the data. Since the IQR removes the lowest and highest 25% of the data, it is not sensitive to outliers. For our data sample, $IQR(\mathcal{S}) = Q3 - Q1 = 2.75 - 1.0 = 1.75$. This is substantially lower than the range as it removed the outlier.

A *five-number summary* is very common and useful for reducing a data distribution to five statistics, namely the minimum, Q1, median, Q3, and maximum. These numbers can be depicted graphically with a *boxplot* as in Figure 9.1. The example boxplot summarizes randomly generated data following a standard normal distribution. Notice that the median does not necessarily

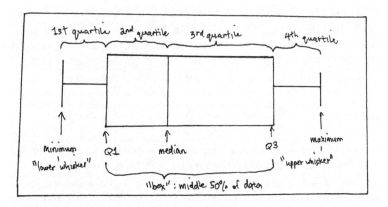

FIGURE 9.1: Boxplot Example

need to place in the middle of the boxplot. Here is a list of the boxplot components:

- The leftmost vertical line (*lower whisker*) represents the minimum

- The left side of the box represents $Q1$

- The range from the minimum to $Q1$ represents the lowest 25% of the data, or first quartile

- The left portion of the box represents the second quartile

- The vertical line in the box represents the median

- The right portion of the box represents the third quartile

- The right side of the box represents $Q3$

- The box represents the middle 50% of the data

- The upper whisker represents the maximum

- The range from $Q3$ to the maximum represents the fourth quartile

Another variant of the boxplot draws whiskers at a distance of $1.5 \times IQR$ from $Q1$ and $Q3$. Data beyond these whiskers are considered outliers and plotted as points. This is done in Python, for example. The code to generate the data and boxplot are shown in the following code snippet. The parameter whis is set to a high level to create the boxplot with whiskers at the minimum and maximum as in the first approach.

```
z = stats.norm.rvs(size=200)
sns.boxplot(z, orient='v', whis=1000)
```

The *sample standard deviation* is a measure of spread based on the deviation of each data point x_i from the sample mean; the deviations are computed as $x_i - \bar{x}$. For aggregating the deviations, the choice is made to add the sum of squared deviations $(x_i - \bar{x})^2$; adding them directly would produce zero due to sign cancellations. The sample standard deviation is denoted s and defined as

$$s = \sqrt{\frac{\sum_{i=1}^{n}(x_i - \bar{x})^2}{n-1}}$$

The sample standard deviation is an estimate of the *population standard deviation*, denoted σ. Squaring the sample standard deviation produces the *sample variance* denoted s^2 and defined as

$$s^2 = \frac{\sum_{i=1}^{n}(x_i - \bar{x})^2}{n-1}$$

The variance is particularly useful when combining the spread of several variables, as variances can be directly added; this is not the case with standard deviations, as they involve square roots. The divisor $n-1$ produces an unbiased estimate of s^2. This means that if we draw random samples from the population, the average value of s^2 taken over the random samples will be equal to σ^2, the *population variance*.

Comovement

It is particularly useful in prediction to understand how two variables move together. Given equally sized samples of two random variables X and Y, we can compute statistics which measure their comovement. The *sample covariance* between X and Y can be denoted s_{xy} and computed in a similar way to the sample variance. The sample variance uses a squared term $(x_i - \bar{x})^2$; for the case of covariance, we replace this with $(x_i - \bar{x})(y_i - \bar{y})$ where $\bar{x}$ and $\bar{y}$ represent the sample means of x and y, respectively. The sample covariance may then be defined as

$$s_{xy} = \frac{\sum_{i=1}^{n}(x_i - \bar{x})(y_i - \bar{y})}{n-1}$$

The sample covariance s_{xy} is an estimate of the *population covariance* denoted σ_{xy}. The sign and magnitude of the covariance indicates the average tendency of the variables to move together. A large, positive covariance indicates that the variables move in the same direction, while a large, negative covariance indicates movement in the opposite direction. A covariance of zero indicates no relationship between the variables.

For two predictors X and Z, we may compute their covariances with a target Y as s_{xy} and s_{zy}, respectively. As the covariances are not adjusted for the spreads of the underlying variables, it is not possible to directly compare them, however. It will therefore be useful to normalize, or scale, the covariance by the product of the variables' standard deviations. If s_x and s_y are the sample standard deviations of X and Y, respectively, then the *sample correlation* between X and Y, denoted r_{xy}, is defined as

$$r_{xy} = \frac{s_{xy}}{s_x s_y}$$

Given this scaling, the sample correlation will fall in the range $[-1, 1]$. The interpretation of correlation is identical to covariance: a higher, positive value indicates that the variables move in the same direction, on average, while a negative value indicates movement in opposite directions. A correlation of zero indicates no relationship. Correlation estimates between predictors and a target will be essential in identifying good predictor candidates for a model.

We have now seen the essential statistics needed to better understand our data: frequency counts, measures of central tendency and dispersion, and measures of comovement. Given one or more datasets, these statistics can be computed for better insight. Pairing statistics with graphics provides for robust exploratory data analysis. We will discuss this topic in detail further into the book.

9.2   Inferential Statistics

The topic of *inferential statistics* is vast. For our purposes, we will review key ideas and specific examples. Suppose that we set out to provide an *alternative hypothesis* (denoted H_A) about a population parameter θ. The current belief is called the *null hypothesis*, denoted H_0. The current belief might be that θ is equal to some value θ_0, while the alternative belief is that θ differs from θ_0.[1] This can be stated formally as

$$H_0 : \theta = \theta_0$$

$$H_A : \theta \neq \theta_0$$

For example, we might hypothesize that the mean batting average of a baseball player is 0.300. In this case, the parameter of interest is the population mean μ, the null hypothesis is $H_0 : \mu = 0.300$, and the alternative is $H_A : \mu \neq 0.300$.

To provide evidence to reject the null hypothesis H_0, we can collect a sample of data from the population and compute an estimate called the *test*

---

[1]Note that other hypothesis structures are possible, such as testing $\theta > \theta_0$ versus $\theta \leq \theta_0$.

statistic. The fundamental challenge is that each sample might produce a slightly different estimate due to sampling error. Increasing the sample size will reduce this sampling error, but including more data is not always possible. We will therefore need to account for the variability of the estimate when making a decision.

The estimate will follow a distribution called the *sampling distribution*, and for large samples ($n \geq 30$) this will increasingly resemble a normal distribution due to the central limit theorem. In data science, one can expect the samples to nearly always surpass 30 by a large margin. After standardizing the estimate, it will follow a standard normal distribution, and probabilities can be computed with software (the modern approach) or found in tables (the earlier approach).

It will be of interest to compute a probability that answers this question: "What is the probability of observing this value of the test statistic or something more extreme assuming that H_0 is true?" The "more extreme" statement is determined from H_A. This probability is called the *p-value*. When the p-value is "sufficiently small," we can reject H_0; otherwise we fail to reject H_0. The benefit of this approach is that it is driven by the data. We will discuss the choice of threshold for making the decision next, but let's first consider a small example where

$$H_0 : \theta \geq 0$$

$$H_A : \theta < 0$$

and the test statistic is –0.25. The question is then: "What is the probability of a test statistic of –0.25 or less assuming that $\theta \geq 0$?" Intuitively, if this probability (or p-value) is very large, this means that H_0 is likely true. However, if the p-value is very small, then there is evidence to reject H_0 in favor of H_A.

There are two possible errors that can be made with this approach. If we reject the null when it is true, this is called a *Type I error*. If we fail to reject the null when it is false, this is called a *Type II error*. Since uncertainty cannot be eliminated, the strategy is to accept some probability of making a Type I error called the *significance level*, denoted α. The most common value for α is 0.05, which amounts to potential error in 1 out of 20 decisions. A less stringent value of 0.10 (1 out of 10) and a more stringent value of 0.01 (1 out of 100) is also common. These numbers are not magical and should be taken with a grain of salt when results are borderline. A two-way table of decision versus actual (or truth) is shown below. For example, the decision to reject H_0 (accepting H_A) when H_A is true results in a correct decision.

decision/actual	H_0	H_A
H_0	correct	Type II error
H_A	Type I error	correct

We now have all of the necessary ingredients for a hypothesis test. We can state H_0 and H_A for a given parameter.[2] We set a significance level α and a sample size n. A statistic is selected to estimate the value of the parameter. Then given the data, the test statistic is computed. For a large sample, the test statistic will have an approximate standard normal distribution after standardization. The p-value can be computed and compared against α. If the p-value is less than α, then H_0 is rejected, as the probability of making a Type I error is sufficiently small. If the p-value is greater than or equal to α, then we fail to reject H_0.

9.2.1   One-Sample Test of the Mean

Next, we will conduct a specific test on the population mean μ. The hypothesis test will look like this:

$$H_0 : \mu = \mu_0$$
$$H_A : \mu \neq \mu_0$$

Given the alternative hypothesis, this is a *two-tailed test* where we must consider the parameter taking a value less than μ_0 or greater than μ_0. When we compute the p-value, we must calculate the probability of a test statistic taking its value or something more extreme in either direction. For example, if the test statistic $t = -1$, then the p-value is the probability that $t < -1$ or $t > 1$ assuming that H_0 is true. This can be written $P(|t| > 1)$.

The sample mean $\bar{X}$ will be used to estimate μ. As its distribution will depend on the sample size n, we will denote this random variable as $\bar{X}_n$. The central limit theorem makes this statement about the sampling distribution of the mean:

As $n \to \infty$,

$$\sqrt{n}\left(\frac{\bar{X}_n - \mu}{\sigma}\right) \to \mathcal{N}(0, 1)$$

Before standardization, this says that as n increases, $\bar{X}$ is approximately normally distributed with mean μ and standard deviation $\sigma/\sqrt{n}$. The quantity $\sigma/\sqrt{n}$ is called the *standard error* (SE) of the sample mean. In general, the SE of a statistic is the standard deviation of its sampling distribution. As the sample size increases, the standard error decreases, with $\bar{X}_n$ approaching the population mean μ.

One complication is that we generally won't know the population standard deviation σ. In such a case, we can replace it with the sample standard deviation s. Now let's look at the quantity

$$\frac{\bar{X}_n - \mu}{s/\sqrt{n}}$$

---

[2]It is also possible to test functions of multiple parameters, such as the difference $\theta_1 - \theta_2$.

This is the standardized form of $\bar{X}_n$ which measures its distance from μ in units of standard error. For $n >= 30$, this quantity approximately follows the standard normal distribution, and it otherwise follows a *t-distribution*. The t-distribution has more probability in the higher and lower values, or *tails*, and its shape depends on a parameter called the *degrees of freedom*, denoted ν. Like the standard normal, the t-distribution is symmetric about zero. Figure 9.2 shows the probability density functions of different t-distributions; the case where $\nu = +\infty$ depicts the standard normal distribution. Notice that the standard normal distribution has much less probability in the tails where $P(X < -2)$ and $P(X > 2)$.

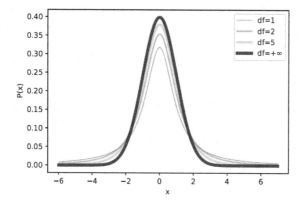

FIGURE 9.2: Student's t-distribution

For our purposes, we will assume $n > 30$ and treat $z = \frac{\bar{X}_n - \mu}{s/\sqrt{n}}$ as a standard normal. Given the data, we can compute the sample mean, the sample standard deviation, and the test statistic z. The p-value is computed under the two-tailed test, which means it reflects the probability in both tails. Lastly, the p-value is compared against α to make a decision.

Let's consider a specific example where we test if the population mean is statistically different from zero:

$$H_0 : \mu = 0$$

$$H_A : \mu \neq 0$$

The parameters are $n = 1000$ and $\alpha = 0.5$, and the sample statistics (computed from the data sample) are $\bar{X}_{1000} = -0.5$, and $s = 10$. From this sample, the mean is different from zero, but the question is whether the difference is statistically significant. We compute the test statistic as

$$z = \frac{\bar{X}_n - \mu}{s/\sqrt{n}}$$
$$= \frac{-0.5 - 0}{10/\sqrt{1000}}$$
$$\approx -1.581$$

Next, we can compute the probability of this value or lower under the standard normal distribution. This represents the left-tailed probability. We can then double this value to include the right-tailed probability. From Python, we can call the function `stats.norm.cdf(z)`, which computes the left tail using the cumulative distribution function of a standard normal. Doubling the probability gives a p-value of approximately 0.1138. As the p-value is greater than $\alpha = 0.05$, we cannot reject H_0 as the risk of making a Type I error would be too large. Thus, based on the data sample, the mean μ is not statistically different from zero.

Python code is shown below for experimenting with different values and measuring the impact on the conclusion. For example, changing the sample mean from –0.5 to –1.0 will produce a p-value of 0.00157 and lead to rejecting H_0.

```
from scipy import stats

n     = 1000
alpha = 0.05
xbar  = -0.5
mu    = 0
s     = 10

test_stat = (xbar - mu) / (s/np.sqrt(n))

# p-value for two-tailed test
pval = stats.norm.cdf(test_stat) * 2

print('test-stat:', round(test_stat,4))
print('p-value (two-tailed):', round(pval,4))

OUT:

test-stat: -1.5811
p-value (two-tailed): 0.1138
```

One-Sample t-Test Example

When we study regression in later chapters, we will see parameter estimates organized as in Table 9.1. The table summarizes the contribution and importance of predictors in the model. For background, this model predicts Y using a linear equation

$$\hat{Y} = \hat{\beta}_0 \times Intercept + \hat{\beta}_1 \times alcohol + \hat{\beta}_1 \times pH$$

The quantity $\hat{Y}$ denotes a prediction. The values $\hat{\beta}_0$, $\hat{\beta}_1$, and $\hat{\beta}_2$ are coefficient estimates (called `coef`) in the table. It is important to understand if these estimates are different from zero, as this would suggest their associated predictors are important. While the estimates may have values like 1 or 2, it is important to take their standard error into account when deciding if they are significantly different from zero.

Let's walk through the table to make sense of it. The rows `alcohol` and `pH` represent different predictors to be tested. The `Intercept` is a constant added to the model; we can treat this as a third parameter estimate to be tested. The hypotheses for `alcohol`, for example, are $H_0 : \beta_1 = 0$ versus $H_A : \beta_1 \neq 0$. This is a two-tailed test to determine if β_1 is significantly different from zero. The data sample is used to compute the estimate of $\hat{\beta}_1 = 0.3093$.

As a different sample will produce a different estimate, we will need to know the standard error of $\hat{\beta}_1$; this is shown to be 0.009 in the `std err` column. Next, a test statistic needs to be computed. It will have the form (estimate – hypothesized value)/standard error. In this case, the test statistic is a *t-statistic* (shown as column t) and the hypothesized value is 0 under H_0. The test statistic reduces to t = estimate/standard error, which equates to 33.207 when plugging in non-rounded values.

Given t, the p-value of a two-tailed test is computed from the t-distribution with appropriate degrees of freedom ν. In this case, ν is equal to the number of observations minus the number of predictors. The large value t=33.207 produces a very small p-value, shown as 0.000. This is sufficient to know that the p-value is less than $\alpha = 0.05$, which leads us to reject H_0 at this level and conclude that β_1 is significantly different from zero. We reach the same conclusion for `Intercept` and `pH`: at level $\alpha = 0.05$, their coefficient estimates are significantly different from zero.

TABLE 9.1: Regression Parameter Estimate Table

	coef	std err	t	$P > \lvert t \rvert$	[0.025	0.975]
Intercept	1.744	0.250	6.966	0.000	1.252	2.233
alcohol	0.3093	0.009	33.207	0.000	0.291	0.328
pH	0.2770	0.076	3.649	0.000	0.128	0.426

To understand the rightmost columns with headers [`0.025` and `0.975`], we will review confidence intervals in the next section. We will then return to the table to round out the discussion.

9.2.2 Confidence Intervals

When we computed the sample mean, this was a *point estimate* of the population parameter μ. The *confidence interval* (CI) is a range that supplements the point estimate with information about its uncertainty. As an example, I used to visit my mother on Long Island, NY from Manhattan. She would ask what time I would arrive, and I would give a time like noon. Due to traffic on the expressway, we noticed that I tended to arrive in a range that was $\pm$ 30 minutes of the target time. Eventually, we would form an interval for planning my arrival: if I said noon, we would plan for 11:30 am–12:30 pm. The interval was more likely to be correct, and we both were happier.

The confidence interval can be regarded from a frequentist point of view like this: we can imagine taking a large number of samples from the population. For each sample, we can construct a confidence interval with a desired confidence level, say 95% (this is most common). The fraction of confidence intervals containing the true parameter will approach 95% as the number of samples increases.

For a parameter θ, the interval is centered at the point estimate $\hat{\theta}$, and a multiple of the standard error is subtracted and added to form the lower limit and upper limit, respectively. The multiple is called the *critical value* (cv) and it is chosen to construct the CI with a desired confidence level. Specifically, it is the value that puts probability $1 - \alpha$ in the interval and $\alpha/2$ in each tail. For $\alpha = 0.05$, this would place 0.95 probability in the interval and 0.025 in each tail. The confidence interval would look like

$$\hat{\theta} \pm cv \times SE$$

or equivalently

$$(\hat{\theta} - cv \times SE, \quad \hat{\theta} + cv \times SE)$$

For the travel example, $\hat{\theta}$ is noon and $cv \times SE$ is 30 minutes. The interval is then (11:30 am, 12:30 pm).

Next, let's consider this specific case: we want to construct a 95% confidence interval for the population mean μ, the population standard deviation σ is unknown, and we have drawn a small sample of size $n = 20$. We can use $\bar{x}$ as a point estimate for μ and the sample standard deviation s as a point estimate for σ. Suppose $\bar{x} = 15$ and $s = 2$.

Since the sample is small, the sampling distribution of $\bar{X}$ follows a t-distribution with $\nu = n - 1 = 19$ degrees of freedom. Then the confidence interval can be expressed like this:

$$\left(\bar{x} - t_{n-1, \alpha/2} \cdot \frac{s}{\sqrt{n}}, \quad \bar{x} + t_{n-1, \alpha/2} \cdot \frac{s}{\sqrt{n}}\right)$$

In this formula, we substituted $\hat{\theta} = \bar{x}$, $SE = \frac{s}{\sqrt{n}}$ and $cv = t_{n-1,\alpha/2}$. The last term is the critical value for a t-distribution with given ν and probability $\alpha/2$ in the right tail; it can be computed with software or found in a t-distribution table.

To determine the critical value using Python, we can run the code below. The function `scipy.stats.t.ppf()` will calculate the critical value of the t-distribution with two parameters. The first parameter represents the probability below the critical value, which needs to be 97.5%. This is because there needs to be 2.5% probability in the right tail of the 95% confidence interval. The second parameter represents the degrees of freedom, which is 19.

```
import scipy.stats

# find critical value of t-distribution
scipy.stats.t.ppf(q=1-.05/2, df=19)

OUTPUT:
2.093024054408263
```

Let's substitute the values into the confidence interval expression:

$$(15 - 2.093 \cdot \frac{2}{\sqrt{20}}, \ 15 + 2.093 \cdot \frac{2}{\sqrt{20}})$$

Simplifying gives this 95% confidence interval for the population mean:

$$(14.0640, 15.9360)$$

A confidence interval can be used for conducting a two-tailed hypothesis test. If the hypothesized value from H_0 does not fall in the confidence interval, then we can reject H_0 at significance level α. Reproducing our parameter estimates in Table 9.2, the two rightmost columns with headers [0.025 and 0.975] show 95% confidence interval lower limits and upper limits, respectively. For example, the 95% confidence interval for the parameter on `alcohol` is (0.291, 0.328). The null hypothesis assumes a parameter value of zero for each parameter test. Since zero is not contained in the confidence intervals, the null can be rejected in each test.

TABLE 9.2: Regression Parameter Estimate Table

	coef	std err	t	$P > |t|$	[0.025	0.975]
Intercept	1.744	0.250	6.966	0.000	1.252	2.233
alcohol	0.3093	0.009	33.207	0.000	0.291	0.328
pH	0.2770	0.076	3.649	0.000	0.128	0.426

9.3   Chapter Summary

In studying a population, it generally is not possible to take a full census. We draw samples and compute statistics with the aim that the statistic can tell us something meaningful about the population. Descriptive statistics help us understand the population. Inferential statistics allow us to draw conclusions about the population based on the sample.

When we have a data sample, there are several statistics we can compute to understand its distribution. The frequency counts tell us the number of times each value is taken. We studied several measures of central tendency: the mean, median, and mode. The mode is the most frequent value. It is simple to explain, but it is purely frequency based. The median is the 50th percentile of the data. Since it ranks the data, it is not influenced by outliers and it better identifies the center of the data. The average or sample mean of the data is the sum of the data divided by the number of observations. It is a point estimate of the population mean. As each data point contributes equally to the average, it can be heavily influenced by outliers.

We studied several measures of dispersion or spread in data. The range is the difference between the maximum and minimum. It is simple to calculate, but it can be influenced by outliers. The interquartile range is the range covered by the middle 50% of the sorted data. By trimming the lowest and highest 25%, the outliers are removed. The five-number summary is useful for reducing a data distribution to a handful of values: the minimum, 25th percentile, median, 75th percentile, and maximum. The boxplot is useful for graphically depicting the distribution with the five-number summary. The sample standard deviation measures the spread based on the deviation of each data point from the sample mean. The sample variance is the square of the standard deviation.

The sample covariance measures the average tendency of two variables to move together. A large, positive covariance indicates that both variables move in the same direction, while a large, negative covariance indicates movement in the opposite direction. If the covariance is zero, this indicates no relationship between the variables. The covariance can be divided by the product of the variables' sample standard deviations to produce the sample correlation. Correlations are typically reported in favor of covariances as they fall in the range $[-1, 1]$.

A point estimate is a single value that approximates a parameter, such as the sample mean $\bar{x}$ estimating the population mean μ. A confidence interval is a range which supplements the point estimate with a variability measurement. It can also be used for conducting a two-tailed hypothesis test. We reviewed the construction of a confidence interval for the population mean.

We outlined the formal process of hypothesis testing. This uses a data sample to infer information about a population parameter. One-sample tests were reviewed for building a conceptual understanding. The various tests will differ, as the sampling distribution will depend on the size of the data sample, the parameter being tested (e.g., the population mean), and the availability of the population standard deviation. We examined a regression parameter estimate table to understand what the output looks like and what it means. The table contains quantities which are essential to any hypothesis test: point estimates, standard errors, test statistics, and p-values. More generally, hypothesis tests are routinely computed with software. Manual calculations are not usually necessary, but it is essential to understand what is being done and what assumptions are being made.

There are several other common hypothesis tests, such as the one-sample test of proportion and the two-sample t-test. The interested reader should explore these tests, but they won't be necessary for advancing through this book. Each of these tests will use the ideas of setting up a null and alternative hypothesis, working with a sample statistic, and a computing standard error, test statistic, and p-value. For a broad, excellent resource on statistical methods, please see [30].

9.4 Exercises

Exercises with solutions are marked with Ⓢ. Solutions can be found in the `book_materials` folder of the course repo at this location:

https://github.com/PredictioNN/intro_data_science_course/

1. Ⓢ How is descriptive statistics different from inferential statistics?

2. Ⓢ The five-number summary is something of a compromise between a single statistic and a histogram. Explain what this means.

3. Ⓢ When is the median preferred over the mean as a measure of central tendency?

4. Ⓢ For a data distribution that is skewed to the right, what does this tell us about the relationship between the mean and median?

5. Consider a dataset consisting of one million ones. What is its standard deviation?

6. Explain when the interquartile range will be preferred over the range as a measure of spread.

7. Ⓢ Explain how correlation is different from covariance.

8. A random variable X follows a standard normal distribution. A

random variable Y follows a t-distribution with 5 degrees of freedom. What is the probability that X is greater than 2? What is the probability that Y is greater than 2?

9. Ⓢ A marketer observes a response rate of 40% from campaign A and 50% from campaign B. Explain why hypothesis testing is useful in determining if these rates are different.

10. What is sampling error? How does increasing the sample size affect the sampling error?

11. Ⓢ Explain what the p-value represents.

12. Ⓢ What is the interpretation of a confidence interval?

13. Ⓢ True or False: A 95% confidence level means that for a given interval, there is a 95% probability that the population parameter falls within the interval.

14. A data scientist builds a model with a predictor with mean value 5. In a recent batch of 1000 observations, the mean was 5.5 and the sample standard deviation was 2. At significance level 0.05, is the new mean different from the old mean? Provide your reasoning.

15. Ⓢ Construct a 95% confidence interval for the mean based on a sample of size 500 with sample mean 10 and sample standard deviation 20.

16. Construct a 99% confidence interval for the mean based on a sample of size 500 with sample mean 10 and sample standard deviation 20.

10

Data Transformation

We have ingested and processed raw data into cleansed data that promises to be useful in further analysis. The end goal might be visualizing data for review by leadership, producing quantitative summaries to share with customers, or a machine learning model for predicting an outcome. There is a large step that is missing, however, and that is maximizing the signal in the cleansed data. Specifically, we might want to know how two variables are correlated. It might be important to understand if one or more variables are trending over time (that is, if their means are increasing or decreasing).

Unfortunately, there are factors that get in the way of easily discerning a signal in data, and chief among them are noise, scale, and data representation. We will start by exploring transformations, or transforms, for handling these challenges. In the interest of space, we will cover a selection that is common and useful.

It is also possible to create new predictors which may produce insights and improve models. This activity can be a fun and creative process. We will consider some examples for illustration.

10.1   Transforms for Treating Noise

Many variables are subject to different kinds of noise, or error. *Measurement error* can result from random error in a system, such as electrical disturbances. *Human error* can result in transmitting incorrect data, such as permuting digits or entering an extra zero. This is also called a *fat finger error*, which originated in financial trading markets. Sampling error can result from a data sample that fails to accurately represent the population under study.

When we think about a process that fluctuates wildly and is subject to signals and noise, the stock market might come to mind. Stock prices move for reasons including fundamentals, economic events, and randomness at different time scales. Figure 10.1 shows an example of a simulated stock price over time. We see many peaks, valleys, and jagged movements. In real-world stock data, the jagged movements are often induced by noisy market microstructure. This noise can make it difficult to identify trends, and it can render the data series less useful as a predictor in a model.

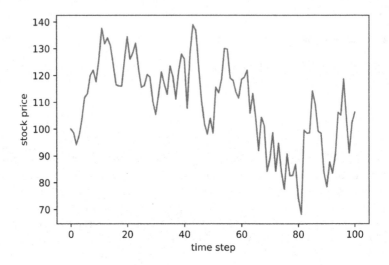

FIGURE 10.1: Stock Trajectory

10.1.1 Moving Average

A common technique for denoising the data is to smooth it with a moving average. A twenty-period moving average, for example, would compute the average price over rolling twenty-day windows. Typically, the first twenty days comprise the first window, days two to twenty-one comprise the second window, and so on.

The `pandas` module provides a function for easily doing this, so we store the prices in a dataframe and call the `rolling()` function like this:

```
steps = 100

# construct a dataframe, passing the data as a dictionary
df = pd.DataFrame({'time':range(steps), 'price':prices})

# append the moving average column
df['ma20'] = df.price.rolling(20).mean()
```

The stock price and moving average trajectories are shown in Figure 10.2. The moving average curve is much smoother than the prices, which helps to see the trend. The moving average exhibits a lag, responding slower to movements. This is apparent where the stock price sharply dropped around time step 60, while the moving average fell much more slowly. Finally, the moving average curve does not begin until time step 20, which matches the number of points used in the average. In general, a moving average with a longer term will require more points to begin, and it will induce more smoothing over the raw data.

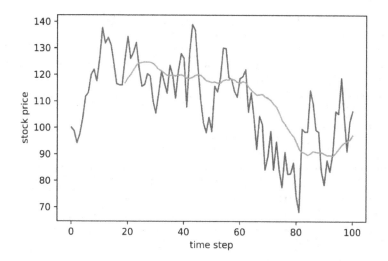

FIGURE 10.2: Stock with 20-Day Moving Average

10.1.2   Limiting Extreme Values

Two popular approaches for limiting the effects of extreme values, which may be produced by noise, are *winsorizing* and *trimming*. Care needs to be taken with these approaches, as outliers may also be legitimate data.

In winsorizing, outliers are set to given percentiles of the data. For example, 80% winsorization would leave the middle 80% of the data untouched. Values below the 10th percentile would be floored to the 10th percentile, and values above the 90th percentile would be capped at the 90th percentile. The code below winsorizes data containing outliers. The bottom 10% of the data is floored at 1 (the 10th percentile), while the top 10% is capped at 9 (the 90th percentile).

```
from scipy.stats.mstats import winsorize

data_win = [-1000, 1,2,3,4,5,6,7,8,9,1000]
winsorize(data_win, limits=[0.1, 0.1])

OUT:
masked_array(data=[1, 1, 2, 3, 4, 5, 6, 7, 8, 9, 9],
             mask=False, fill_value=999999)
```

Winsorizing can be helpful in measuring a statistic that is robust to outliers. If we wanted to understand the center of the data, we might compute the mean. The mean of the non-outlier values [1,2,3,4,5,6,7,8,9] is 5. For this case, winsorizing the data and computing its mean will match this value. Had we included the outliers in the calculation, they would have exerted a large influence and produced a mean of approximately 4.09.

Trimming data is different from winsorizing, as it discards outliers. A 20% trimmed mean would remove the lowest 10% and highest 10% of the data and compute the mean. Continuing with our example data, the 20% trimmed mean would also be 5. The code to implement this is the following:

```
from scipy import stats
stats.trim_mean(data_win, 0.2)
```

In this section, we looked at transforms for filtering noisy measurements from data. There are many other useful filters, such as the low-pass filter, Kalman filter, and Butterworth filter. The analyst will need to decide on the best transform and its required parameter values. This will depend on the data and the use case. Visualization and automated evaluation can help. For an excellent treatment of digital filters, please see [31].

10.2 Transforms for Treating Scale

10.2.1 Order of Magnitude and Use of Logarithm

When we think about scale, let us think about a variable X that takes values in [0,10], and a second variable Y that generally takes values in [10, 20], but occasionally takes much larger values such as 10,000. A value's *order of magnitude* is that value expressed as a power of 10. For example, 10,000 expressed as a power of 10 is 4. It can be written in scientific notation as 1×10^4. Next, let's think about differences in orders of magnitude between two numbers. The difference in order of magnitude between 100 (or 1×10^2) and 10,000 (or 1×10^4) is two. Large differences in orders of magnitude can cause trouble when visualizing data and when fitting certain models to data. Let us consider this small example, where x and y each contain lists of values:

```
x = [0,1,2,3,4,5]
y = [10,10,10,20,20,10000]
```

We are expecting the y-value of 10,000 to cause trouble. Figure 10.3 graphs this data. As we expected, the 10,000 makes the smaller y values appear as zero or nearly zero. Depending on the context, this can be problematic. If the question were: how many fatalities were there in a given year, and y represented these fatalities, then an answer of zero is very different from one, ten, or twenty.

Since we are trying to better understand all of this data, we are looking for a way to better represent it. A common technique when facing values with different orders of magnitude is to take the logarithm of the data using

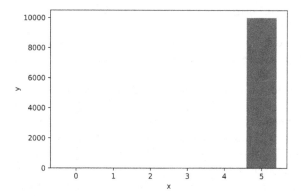

FIGURE 10.3: x versus y

base 10. This is consistent with measuring order of magnitude, which means that a value of 10 will have a logarithm of 1, and a value of 10,000 will have a logarithm of 4. Figure 10.4 demonstrates that the log(y) transformation better represents the entirety of the data.

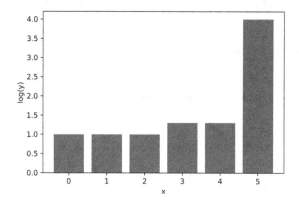

FIGURE 10.4: x versus log(y)

10.2.2 Standardization

In this section, we will spend more time on standardization, which we encountered when working with random variables following a normal distribution. Imagine two different language classes that took the same exam. The grades from each class roughly followed a normal distribution with the same standard deviation of 2. In class 1, the mean was 90, while the mean for the second class was 75. Suppose a student in each class scored a 90.

Which student did better?

The student in class 1 scored at the mean, while the student in class 2 performed remarkably better. To make a fair comparison, we standardize the scores by computing z-scores according to $z = (x - \mu)/\sigma$. A grade of 90 in class 1 equates to a z-score of $(90 - 90)/2 = 0$, while the grade in class 2 is equivalent to a z-score of $(90 - 75)/2 = 7.5$. Note that a z-score of 7.5 is extremely high, recalling that 99.7% of the time, a standard normally distributed random variable will have a z-score between –3 and 3.

In this small example, we standardized one value from each class. In data science, we will often standardize one or more columns of data, where each column is a variable. This places each variable on a similar scale and retains outliers. Standardization in Python is commonly done using the **sklearn** module. The function **StandardScaler()** can be used to standardize one or more columns. Here is a code example for illustration:

```
from sklearn.preprocessing import StandardScaler
import numpy as np

# create an array with four rows and two columns (variables)
data = np.array([[4, 4], [6, 0], [0, 2], [1, 1]])

# set up the scaler
scaler = StandardScaler()

# compute the z-scores
scaled_data = scaler.fit_transform(data)

print(data)
print('')
print(scaled_data)

OUT:
[[4 4]
 [6 0]
 [0 2]
 [1 1]]

[[ 0.52414242  1.52127766]
 [ 1.36277029 -1.18321596]
 [-1.15311332  0.16903085]
 [-0.73379939 -0.50709255]]
```

By default, the scaler computes the mean and standard deviation of each column and produces the z-scores. For this data, each column has values in a range of $[-3, 3]$, which is common for a variable with a standard normal distribution. Additionally, each column will have a mean of zero and standard deviation of one. We can compute the means and standard deviations of each column to check one of the z-scores by "hand."

```
data.mean(axis=0)
OUT: [2.75 1.75]

data.std(axis=0)
OUT: [2.384848  1.47901995]
```

The `axis=0` parameter allows for selecting across rows when computing the statistic. For the value of 4, its z-score is then $(4-2.75)/2.384848 = 0.52414242$ which matches the `scaled_data` entry.

10.2.3 Normalization

Another common approach to scaling data is *normalization*. This approach squeezes the data into a range which is commonly selected to be $[0, 1]$ or $[-1, 1]$. Unlike what the name may suggest, this transform has nothing to do with the normal distribution, and the data does not need to follow a normal distribution. Outliers are floored to the lower bound and capped at the upper bound. This transformation is effective in scaling data, but when the data is approximately normally distributed, standardization is recommended for retaining symmetry about the origin and for retaining outliers. The formula for normalizing data is the following:

$$x_{norm} = \frac{x - x_{min}}{x_{max} - x_{min}}$$

Here is a code snippet for normalizing the dataset from the previous example. By default, `MinMaxScaler()` uses a lower bound of 0 and an upper bound of 1.

```
from sklearn.preprocessing import MinMaxScaler
import numpy as np

# set up the scaler
scaler = MinMaxScaler()

# normalize the data
norm_data = scaler.fit_transform(data)

print(norm_data)

OUT:
```

```
[[0.66666667 1.          ]
 [1.         0.          ]
 [0.         0.5         ]
 [0.16666667 0.25        ]]
```

For each column, the original minimum and maximum were mapped to zero and one, respectively. As the columns are in the same range [0, 1], scaling is achieved.

10.3 Transforms for Treating Data Representation

Recall the California wildfire news article. We processed the data, tokenizing the document into words. Suppose that we wanted to automatically determine, or classify, the topic of the article based on the frequency of the words. This is a standard task accomplished with a machine learning model, but the model cannot directly accept words, which are strings. We will need to transform the data into an appropriate quantitative form. The type of model (e.g., regression) and its implementation will dictate the precise form. In the next sections, we will review common transformations for representing text and categorical data.

10.3.1 Count Vectorizer

To make progress on the automated topic classification task, we can represent each document as a set of words and their frequencies. The `CountVectorizer()` transform from `sklearn` can be used for this task. Here is a short code example that processes two documents; the first is drawn from the wildfire article we encountered earlier, and the second is about sports. By default, each document will be tokenized into words, and a matrix with documents on the rows and the words on the columns is computed. The element in row i and column j of the matrix will show the number of times that word j appears in document i. The matrix of these counts can later be passed to a machine learning model for topic determination.

```
from sklearn.feature_extraction.text import CountVectorizer

docs= ['The 2021 California wildfire season was
a series of wildfires that burned across the
U.S. state of California.',
        'Golden State Warriors basketball team is
        located in California.']
```

```
countvec = CountVectorizer()

# compute word counts by document
X = countvec.fit_transform(docs)

# print the features, which are words
print(countvec.get_feature_names_out(),'\n')

# matrix of counts, with documents on rows and words on columns
print(X.toarray())
OUT:
['2021' 'across' 'basketball' 'burned' 'california'
'golden' 'in' 'is' 'located' 'of' 'season' 'series'
'state' 'team' 'that' 'the' 'warriors' 'was'
'wildfire' 'wildfires']

[[1 1 0 1 2 0 0 0 0 2 1 1 1 0 1 2 0 1 1 1]
 [0 0 1 0 1 1 1 1 1 0 0 0 1 1 0 0 1 0 0 0]]
```

When we call get_feature_names_out(), this returns the set of words extracted from the text, known as the *vocabulary*. From the matrix, each column representing a word's frequency, or count, in the documents can be used as a predictor.

Let's review the matrix of counts. We see for example that the first document contains the word 'california' twice, while the second document has a count of one (see column 5). Many of the words appear in one document and not the other, such as 'wildfires.' These differences can help with topic classification. In summary, we began with data in text form, and we converted it to a numeric representation for eventual use in a model.

10.3.2 One-Hot Encoding

Categorical data will have discrete values where direct computation on the values does not make sense. Marital status (e.g., 'single', 'married'), education level (e.g., 'high school', 'some college'), and zip code (e.g., 11787) are some examples. As the first two examples are string types, it may be obvious they are not quantitative, while zip code might not be so obvious. For each of these cases, they cannot be averaged, for example. To include these variables in a model, they can be transformed to an amenable representation using *one-hot encoding* (OHE).

Consider this data containing various cat breeds:

```
breeds : ['persian','persian','siamese','himalayan','burmese']
```

To encode this categorical variable, we can form a column for each unique breed: burmese, himalayan, persian, and siamese. Each row will refer to a record, and each element will take value zero or one to indicate the breed. It may look as follows:

record	burmese	himalayan	persian	siamese
1	0	0	1	0
2	0	0	1	0
3	0	0	0	1
4	0	1	0	0
5	1	0	0	0

The first two records are persian, and this is represented by the 1 in the persian column and zeroes in the rest. Written as a vector, record 1 is encoded as [0, 0, 1, 0]. We can condense this further by observing that for a categorical variable with n levels, or unique values, we need only **n-1** columns to represent them. In this case, we will drop the burmese column and update the matrix.

record	himalayan	persian	siamese
1	0	1	0
2	0	1	0
3	0	0	1
4	1	0	0
5	0	0	0

Record 5 takes value 'burmese', and this is reflected by each of the remaining cat breeds taking values of zero. The columns in this matrix are sometimes called *indicator variables* or *dummy variables*. These columns can now be included in modeling techniques such as regression. Next, let's look at a code snippet that begins with the categorical data and converts it to numeric using one-hot encoding.

```
# create a dataframe containing a categorical column
cats = pd.DataFrame(
{'breed':['persian','persian','siamese','himalayan','burmese']}
)

print('--categorical data')
print(cats)

# one hot encoding, dropping the first column by alphabet.
# assign prefix to column names for systematic identification
cats = pd.get_dummies(cats.breed, drop_first=True,
```

```
                        prefix='breed')

print('\n')
print('--one hot encoded categorical data')
print(cats)

OUT:
--categorical data
         breed
0      persian
1      persian
2      siamese
3    himalayan
4      burmese

--dummified categorical data
   breed_himalayan  breed_persian  breed_siamese
0              0              1              0
1              0              1              0
2              0              0              1
3              1              0              0
4              0              0              0
```

10.4   Other Common Methods for Creating Predictors

10.4.1   Binarization

Binarizing data is commonly used to build a variable to reflect if something happened or not (e.g., a user rented a video, a patient recovered). A binarized variable will sometimes be a better predictor than the original variable, as it can condense rare cases where counts are high. For example, in a dataset of movie viewers, it may be that most viewers either watched *Titanic* once or they didn't watch the film. There may be a small number of viewers who watched it ten or more times. To consolidate the information content of this data, it could be useful to binarize the variable into viewers and non-viewers.

A second common use case for binarizing data is when *thresholding* a predicted probability to assign a likely outcome. We will talk more about thresholding later, but essentially if a model predicts that the probability of an outcome is 0.80 (for example), the value will need to be compared to a cutoff to give a yes/no answer.

Mathematically, binarization is a step function. For example, given input x and cutoff c, the function $f(x)$ may be defined:

$$f(x) = \left\{ \begin{array}{ll} 0 & x \le c \\ 1 & x > c \end{array} \right.$$

Here is a code snippet to binarize data using `sklearn`. `Binarizer()` takes a threshold which is zero by default. Values less than or equal to the threshold will be set to zero, and set to one otherwise.

```
from sklearn.preprocessing import Binarizer

bin = Binarizer() # set up the scaler
bin_data = bin.fit_transform(data)

print(data)
print('')
print(bin_data)

OUT:
[[4 4]
 [6 0]
 [0 2]
 [1 1]]

[[1 1]
 [1 0]
 [0 1]
 [1 1]]
```

10.4.2 Discretization

Discretizing a continuous variable into buckets can sometimes yield a more powerful predictor. In particular, it can be effective when the variable has a wide range and it is used in a regression model. The `qcut()` function from `pandas` can discretize data into equal-sized buckets based on ranks or quantiles.

The example below creates 1000 random values drawn from a standard normal distribution. They are multiplied, or *scaled*, by 100 to increase their range. The data is then divided into deciles, where decile 0 is the lowest 10% of data values, decile 1 is the next 10%, and so on. The original data and the deciles are plotted in Figures 10.5(a) and 10.5(b) to show the discretized effect.

```
data = 100*np.random.normal(size=1000)
data.sort()

# discretize into deciles
```

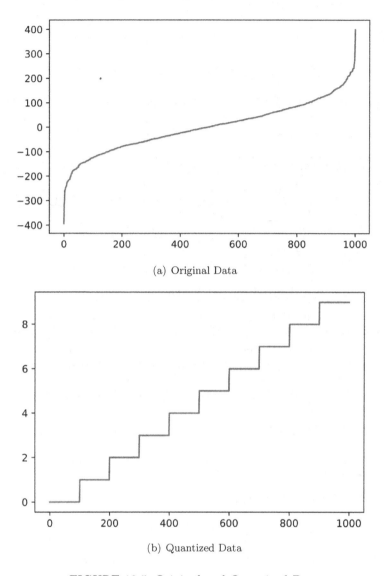

(a) Original Data

(b) Quantized Data

FIGURE 10.5: Original and Quantized Data

```
data_dec = pd.qcut(data, q=10, labels=False)

plt.plot(data)
plt.plot(data_dec)
```

Notice the *sawtooth pattern* in the quantized data. As the data are grouped into deciles, this produces a series of step functions where values are constant within each decile, and then a jump occurs at the next decile. This has the effect of mapping many values to the same output, and narrowing the range of the data to $0, 1, \ldots, 9$.

10.4.3   Additional Common Transformations

There is a wide range of other transformations which may be helpful for engineering useful predictors. The context of the problem can help guide transformation use. The small table below provides some examples.

Transformation	Example
Ratio of variables	Assets to liabilities
Difference of variables	Long-term minus short-term interest rate
Lagged variable	Last month medication dose
Cumulative sum of variable	Total study time
Power of a variable	Square of time

10.4.4   Storing Transformed Data

Creating transformed data can be time consuming. Similar to how the cleaned data was stored, it will be beneficial to store the transformed data to avoid having to reconstruct it. If other permissioned users might find the data useful, then providing access and documentation is recommended. A *feature store* is dedicated storage for such data. Some cloud applications, such as Amazon SageMaker, provide this functionality for making features reusable and easily accessible to permissioned users.

10.5   Chapter Summary

Raw data will generally need to be transformed before it can be useful. Data may be noisy, and we discussed smoothing with a moving average. Extreme values, which sometimes originate from noise, can be removed by winsorizing and trimming. Care should be taken when removing outliers.

Variables are often more useful when they are scaled. It is difficult to visualize data that changes in magnitude. Comparing variables of different magnitudes, or using them in algorithms that measure distance, leads to incorrect conclusions. We studied transformations to handle scale, such as use of the logarithm, standardization, and normalization. When deciding between standardization and normalization, the former is recommended when the data is normally distributed. Normalization will squeeze data into a range such as [0, 1], which removes outliers and symmetry.

Data can be represented in different ways, and the representation used will change its utility. Text data cannot be directly used by a model; it needs to first be converted to a quantitative form. We studied the count vectorizer, which can be used to represent a set of documents as a term-document matrix. One-hot encoding can convert categorical data to a vector indicating the variable's level. Some transformations, such as binarization, work by accentuating certain parts of the data. Discretization converts a continuous variable into discrete buckets, which sometimes yields a powerful predictor.

It takes experimentation to discover the best predictors. Once they are found, it is useful to save the code and save the predictors. Transforming data can be resource intensive, and it is best not to repeat valuable work. Additionally, the transformed data can be used by others when helpful.

The `sklearn` module has many functions that make data transformation easier, and we reviewed small examples. Additional resources on data transformation can be found in the course repo in this folder:

`semester1/week_09_10_transform`

It is helpful to be a "collector" of transforms, and this can be done by looking on the web and elsewhere to find interesting and useful approaches. In the next chapter, we will learn techniques and tools for exploratory data analysis. This will allow us to investigate the cleaned and transformed data.

10.6   Exercises

Exercises with solutions are marked with Ⓢ. Solutions can be found in the `book_materials` folder of the course repo at this location:

https://github.com/PredictioNN/intro_data_science_course/

1. Provide two reasons why transforming raw data can be beneficial.
2. Ⓢ You are deciding between the use of a 5-day moving average and a 30-day moving average. Explain how smoothing and the lag effect will differ between the two.
3. Explain how winsoring and trimming are different.
4. Ⓢ What is the order of magnitude difference between five thousand and two billion?
5. Ⓢ What is the order of magnitude difference between 1/10 and 100,000?
6. Give two instances where large differences in order of magnitude can cause trouble.

7. Ⓢ A student scored a 90 on a test. The class average and standard deviation were 75 and 5, respectively. Calculate the student's z-score.

8. Ⓢ A student scored a 75 on a test. The class average and standard deviation were 75 and 5, respectively. Calculate the student's z-score.

9. Ⓢ Use `numpy` and seed 314 to generate 1000 random values from a uniform distribution. Apply `StandardScaler()` to the data and verify that the standardized data has an approximate mean of zero and standard deviation of one.

10. Use `numpy` to generate a random data array from a standard normal distribution. Apply `MinMaxScaler()` to the data and verify that the normalized data has a minimum of zero and a maximum of one.

11. True or False: A categorical column of data can be directly used by a machine learning model.

12. Ⓢ Given the following list of categorical values, create and print the one-hot encoding matrix. The first level based on alphabet should be dropped.

```
edu_level = ['college','some college','associate',
             'graduate']
```

13. Ⓢ Which of these variables are good candidates for binarization? Select all that apply.

    a) A categorical variable such as employment type
    b) A variable indicating if a patron was satisfied with her meal
    c) The number visits that a user made to a website in the last two days
    d) The number of marathons that a group of randomly selected individuals ran last summer.

14. Following the approach of the discretization section, generate 5000 random values from a normal distribution with mean zero and standard deviation 30. Use the `qcut()` function to discretize the data into 20 buckets. Plot the data to show the sawtooth pattern, where the x-axis is the index of the data points and the y-axis is the assigned bucket.

15. Does the discretized function have an inverse? Explain your answer.

11

Exploratory Data Analysis

One of the first things that should be done with a new dataset is to get a feel for the data. We did some of this during the preprocessing step, when we looked for missing values and extreme observations, and when we transformed variables. Every dataset will be different and might contain surprises; understanding the nuances and the tools to treat them will allow for accurate, insightful analysis.

The two activities we will explore are statistical summarization and visualization of the data. As the datasets grow larger, it can be difficult or impossible to review all of the data points, and this is where statistics will help. Statistical summaries, such as computing means and percentiles, can provide a quick overview of the distribution of one or more variables. Aggregations can summarize data along one or more axes, such as computing average daily measurements on high-frequency data. We will see examples of statistical applications in this chapter's exercises.

Properly done graphics will lead the eye to important features of the data, prompting further investigations. Python includes several modules for visualization, with `matplotlib` and `seaborn` among the most popular. We will show several examples of `seaborn` plots to give a sense of its power and ease. `Seaborn` is based on `matplotlib` and it provides a high-level interface for drawing attractive and informative statistical graphics. For customizing `seaborn` graphics, there are times when it is necessary to include elements from `matplotlib`. The space in this book does not allow for a comprehensive treatment of the modules, but the examples to follow will serve as a starting point. For an introduction to `seaborn`, please see [32]. To get started with `matplotlib`, see [33]. Both of these packages will be pre-installed with Anaconda.

Exploratory data analysis (EDA) can be done on variables individually, of course, but many of the interesting insights will come when we consider how one variable changes, or co-varies, with another. This speaks to the importance of understanding relationships between variables, which is essential for prediction. To drive the point home, consider that each time a variable is selected, that data is taken out of context from a large system. If the column of data is a set of student midterm grades, the excluded data might be the students that weren't feeling well that day. For a series of stock prices, the underlying economic regime has been left out. In each case, these excluded variables might be essential to truly understanding the data. By thoroughly studying each variable individually and in groups, it will help build a more complete understanding of what insights the data holds.

When we conduct analysis on a single variable, this is often called a *univariate analysis*. For the case of studying two variables together, it is often called a *bivariate analysis*. Different statistics and graphs are relevant for each, and they are both necessary for a complete study. In the next two sections, we will conduct EDA on a binary target and a continuous target, respectively. The first exercise will consider a synthetic dataset from a bank where the records represent customer check deposits and the target variable is whether the check was fraudulent or not. The second exercise will study population happiness. These exercises should give a good feel for the data exploration process and methods.

11.1   Check Fraud

In this section, we will study a problem faced by banks offering checking accounts to their customers. Check fraud is typically rare, but the total losses can be sizeable and there is great motivation to detect bad checks.

The synthetic dataset for this analysis lives in the course repo here:

`semester1/datasets/check_deposit_fraud.csv`

A Jupyter notebook with code performing EDA lives here:

`semester1/week_11_12_visualize_summarize/check_deposit_eda.ipynb`

We import the data and show the first few rows:

```
df = pd.read_csv('../datasets/check_deposit_fraud.csv')
df.head(3)
```

OUT:

checkid	check_amount	check_signed	is_fraud
0	9076.51	1	0
1	2148.38	1	0
2	604.52	1	0

We can see the schema of the data by issuing:

`df.info()`

OUT:

	Column	Non-Null Count	Dtype
0	checkid	1100 non-null	int64
1	check_amount	1100 non-null	float64
2	check_signed	1100 non-null	int64
3	is_fraud	1100 non-null	int64

A real-world dataset would include additional attributes like bank name, account number, routing number, check number, and so on. Here, attention is drawn to a small number of interesting attributes, as they have exhibited historical predictive ability. For the binary variable check_signed and the target variable is_fraud, we should compute the frequency and percentage of each value. We count the number of fraudulent and non-fraudulent checks with the value_counts() function:

```
df.is_fraud.value_counts()
0    1096
1       4
```

The rows are ordered from highest number of observations to lowest. In this case, there are 1096 rows where is_fraud is 0 (no fraud; the negative label), and 4 rows where is_fraud is 1 (fraud; the positive label). Including the parameter normalize=True will compute their percentages:

```
df.is_fraud.value_counts(normalize=True)
0    0.996364
1    0.003636
```

Over 99% of the checks are legitimate; fraud is fairly rare.

We follow the same procedure for check_signed:

```
df.check_signed.value_counts()
1    1096
0       4
```

```
df.check_signed.value_counts(normalize=True)
1    0.996364
0    0.003636
```

Most of the checks were signed.

For a larger dataset, there might be dozens or more binary variables and categorical variables that can be summarized in this way. It can help to form groupings of variables so they can be processed together. This example forms a binary variable grouping with additional fictitious variables, and computes their frequency distributions in a loop:

```
binary_vars = ['check_signed','check_torn', 'another_binary']

for vars in binary_vars:
    print(df[vars].value_counts())
```

The `check_amount` variable might be a useful predictor. As it is quantitative, let's look at the summary statistics:

```
df.check_amount.describe()
```

```
count      1100.000000
mean       5071.434100
std        2934.500657
min           2.710000
25%        2484.380000
50%        5077.185000
75%        7662.152500
max       10000.000000
```

The data seems to be uniformly distributed with a minimum close to zero and a maximum of 10000. Next, we use the **seaborn** package and plot a histogram to see the full distribution (see Figure 11.1). Since elements from `matplotlib` will be brought into **seaborn**, the former package will be imported as well. Notice the aliases used by convention: **seaborn** is referenced as **sns**, while `matplotlib.pyplot` is referenced as **plt**.

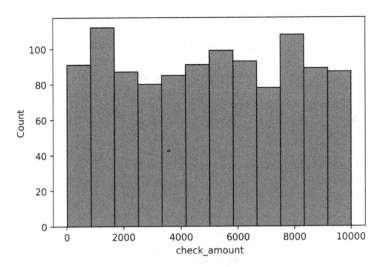

FIGURE 11.1: Histogram of Check Amount

```
import seaborn as sns
import matplotlib.pyplot as plt
```

```
sns.histplot(data=df.check_amount)
```

Tip: the parameter name `data` is optional.

The plot confirms the approximate uniform distribution. The values are all plausible (e.g., there are no negative check amounts).

To change the plot type of a single variable, we would change its name and include the relevant parameters. For example, if we wanted to change from a histogram to a boxplot, the code would change from:

```
sns.histplot(data=df.check_amount)
```

to

```
sns.boxplot(data=df.check_amount)
```

To understand how `check_signed` is distributed for the fraudulent and non-fraudulent checks, we can produce a two-way classification table with the pandas `crosstab()` function:

```
pd.crosstab(index=df['check_signed'], columns=df[is_fraud])
```

```
is_fraud                0       1
check_signed
    0                   1       3
    1                1095       1
```

To read the table, the `check_signed` = 0 row indicates unsigned checks. Reading across that row, of the 4 unsigned checks, 1 was not fraudulent and 3 were fraudulent. Of the 1096 signed checks, 1095 were not fraudulent and 1 was fraudulent.

We can put this in percentage terms for greater clarity, dividing each count by the row sum. Programmatically, this requires adding the parameter `normalize`.

```
pd.crosstab(index=df['check_signed'],
        columns=df[is_fraud], normalize='index')
```

```
is_fraud                0          1
check_signed
    0                0.25       0.75
    1            0.999088   0.000912
```

This says that 75% of the unsigned checks were fraudulent, while only 0.09% of the signed checks were fraudulent. The ratio 0.75 / 0.000912 = 822 strongly suggests that unsigned checks pose a heightened risk of fraud. If we were going to build a predictive model, `check_signed` would be a good candidate predictor.

Next, let's further explore `check_amount` to understand how the amounts vary by the fraud outcome. The scatterplot in Figure 11.2 may clarify their relationship.

```
sns.scatterplot(x='check_amount', y='is_fraud', data=df)
plt.grid() # called from matplotlib
```

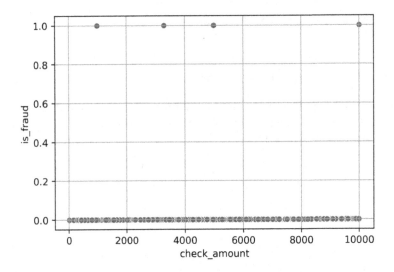

FIGURE 11.2: Check Amount by Fraud Indicator

Gridlines are incorporated from `matplotlib`. For a bivariate plot (also called a *joint plot*), the commonly specified parameters are x, y, and the dataframe `data`. The values of x and y are the variable names entered as strings, while `data` references the dataframe object. Similar to the univariate case, the type of graphic can be changed by swapping the graph type and including the relevant parameters.

Most of the checks were not fraudulent, and their points are plotted on the bottom horizontal grid line. We see the four fraudulent checks on the top horizontal grid line which marks where `is_fraud=1`. The check amounts appear to roughly be valued at 1000, 4500, 5000, and 10000. We can also inspect the fraud records in the dataset:

```
df[df.is_fraud==1]
```

checkid	check_amount	check_signed	is_fraud
101	10000.00	1	1
510	5000.00	0	1
778	1000.00	0	1
901	4615.59	0	1

The table confirms a key observation: For three of the four fraud cases, the check amounts were multiples of 1000. Given this, it makes sense to create a predictor variable which indicates if each check amount is a multiple of 1000. We might name it check_amount_mult_1000.

From our exploratory work, we have uncovered that check_signed and check_amount_mult_1000 are useful predictors of check fraud. In making these insights, we produced statistical summaries, filtered the data, and created plots. We needed to be creative and try different approaches, as the predictors were not immediately apparent.

11.2   World Happiness

The World Happiness Report is an annual survey of the state of global happiness across 117 countries. We will look at data from the 2018 report [34]. Further information can be found here:

https://worldhappiness.report/ed/2018/

and datasets can be found here:

https://www.kaggle.com/unsdsn/world-happiness

The 2018 dataset is saved in the course repo, and we will load it from there.

```
import pandas as pd
import seaborn as sns
import matplotlib.pyplot as plt

df = pd.read_csv('../datasets/Country_Happiness.csv')
df.head(2)
```

OUT:

Country	Happy	WH	WL	Dys	GDP	Social	Health	Free
Finland	7.632	7.695	7.569	2.595	1.305	1.592	0.874	0.681
Norway	7.594	7.657	7.530	2.383	1.4565	1.582	0.861	0.686

Each row represents a unique country, column names are abbreviated, and some columns are suppressed due to margin constraints. We will focus on the happiness score (Happy), which is the target variable, and six factors that contribute to higher life evaluations: economic production (GDP), social support (Social), life expectancy (Health), freedom (Free), generosity (Gener), and perceptions of corruption (Corrupt). Respondents were asked to rate their happiness from 0 (worst possible life) to 10 (best possible life). The hypothesis is that the six factors contribute to the happiness score, either directly (they rise together) or inversely (they move in opposition).

First, let's produce histograms of the happiness score and the factors to understand their distributions. The code is shown below, along with a sample of the histograms.

```
# exclude variables we won't plot
vars_exclude = ['Country','Whisker-high','Whisker-low']

# loop over each variable in dataframe
for var in df.columns:

    # execute block if variable not in exclude list
    if var not in vars_exclude:

        # plot histogram for variable
        sns.histplot(df[var])
        plt.show()
```

The happiness score ranges in value from 2.91 to 7.63, with most scores falling between 4.5 and 6.5 (see Figure 11.3). The lowest score was well above the potential low of zero, and the highest score was well below the potential high of 10.

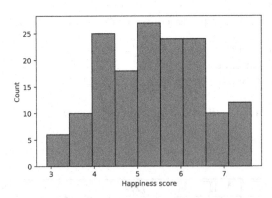

FIGURE 11.3: Distribution of Happiness Scores

The social support factor ranges from 0 to 1.64 and it is left skewed (see Figure 11.4). This means that the distribution is longer to the left of the peak, which occurs at about 1.3.

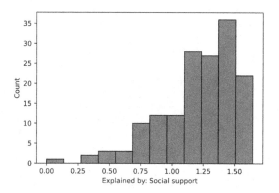

FIGURE 11.4: Distribution of Social Support Factor

The corruption factor ranges from 0 to 0.46. Seventy-five percent of the values are below 0.14, and the distribution is skewed to the right (see Figure 11.5).

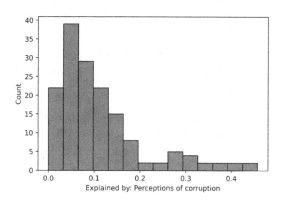

FIGURE 11.5: Distribution of Corruption Factor

Another useful plot for visualizing the distribution of the data is the box-plot. Figure 11.6 shows a boxplot of happiness scores, with the five-number summary shown:

• The maximum appears at the top as a horizontal line (value: 7.63)

• The 75th percentile forms the top of the box (value: 6.17)

• The median, or 50th percentile, is the horizontal line in the box (value: 5.38)

- The 25th percentile forms the bottom of the box (value: 4.45)

- The minimum appears at bottom as a horizontal line (value: 2.91)

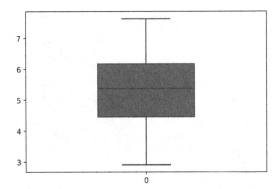

FIGURE 11.6: Boxplot of Happiness Scores

A boxplot provides less detail than the histogram, but it can be easier to interpret. Additionally, several boxplots can be easily compared together.

```
sns.boxplot(data=df['Happiness score'], width=0.4)
```

Next, let's look at bivariate plots of the data to understand the relationships between the factors and the target. We create a scatter plot of happiness versus social support:

```
sns.scatterplot(x='Explained by: Social support',
                y='Happiness score', data=df)
```

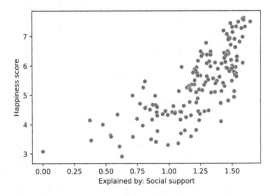

FIGURE 11.7: Happiness Scores versus Social Support

Figure 11.7 shows a direct relationship between social support and happiness; they tend to move together. The correlation is 0.76, which indicates a strong positive (linear) association.

Next, we create a scatterplot of happiness versus corruption:

```
sns.scatterplot(x='Explained by: Perceptions of corruption',
                y='Happiness score', data=df)
```

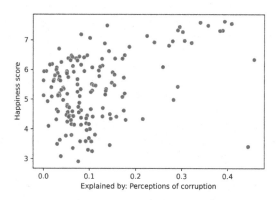

FIGURE 11.8: Happiness Scores versus Corruption

A direct relationship is not apparent here, and there are some outliers to the far right. A correlation of 0.41 suggests a weak direct relationship.

We have separately created univariate plots for the underlying variables and a scatterplot to see their comovements. A jointplot (Figure 11.9) combines the scatterplot with the histograms from each of the underlying, or marginal, variables:

```
sns.jointplot(x='Explained by: Social support',
              y='Happiness score', data=df)
```

Notice that we changed from a scatter plot to a joint plot by changing the word `scatterplot` to `jointplot`. The `seaborn` interface makes this very easy.

Computing correlations is straightforward with **pandas**. All of the pairwise correlations may be computed in a dataframe with the `df.corr()` command. If we would like only the correlations with the target column, we can subset on the column:

```
df.corr()['Happiness score']
```

OUT:

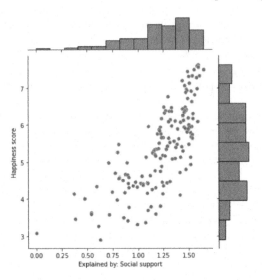

FIGURE 11.9: Jointplot of Happiness Scores versus Corruption

Happiness score	1.000000
Whisker-high	0.999553
Whisker-low	0.999571
Dystopia (1.92) + residual	0.473606
Explained by: GDP per capita	0.807338
Explained by: Social support	0.764656
Explained by: Healthy life expectancy	0.777484
Explained by: Freedom to make life choices	0.562290
Explained by: Generosity	0.142030
Explained by: Perceptions of corruption	0.408650

This shows that the factor with strongest correlation is GDP, while Generosity has the weakest correlation.

Lastly, **seaborn** can graph the joint and marginal distributions for all pairwise relationships and each variable, respectively, with the `pairplot`. The data density of this plot in Figure 11.10 is very high and it is possible to understand overall patterns quickly. Since the column names are long, they are first abbreviated with a passed dictionary of 'old name' : 'new name' pairs.

```
# abbreviate column names
df.rename(columns={
        'Country' : 'country',
        'Happiness score' : 'happy',
        'Whisker-high' : 'wh',
        'Whisker-low' : 'wl',
```

```
        'Dystopia (1.92) + residual' : 'dys',
        'Explained by: GDP per capita' : 'gdp',
        'Explained by: Social support' : 'social',
        'Explained by: Healthy life expectancy' :
                        'exp',
        'Explained by: Freedom to make life choices' :
                        'freedom',
        'Explained by: Generosity' :
                        'generous',
        'Explained by: Perceptions of corruption' :
                        'corrupt'
    }, inplace=True)
```

```
# produce the pairplot with desired size
sns.pairplot(data=df, height=1.25)
```

The histograms appear on the diagonal (for example, the happiness score is at top left) and the scatter plots are off diagonal. The variable names on the x-axis and y-axis can identify a particular scatter plot (for example, row 5, column 1 shows happiness versus GDP). The upward-sloping graph suggests a strong direct relationship.

11.3   Use and Limitations of Summary Statistics

We have studied many variable pairs in this section by graphing scatter plots and computing correlations. More generally, the two common pairings we will study are predictor-target pairs and predictor-predictor pairs. Finding a strong relationship – direct or indirect – between a predictor X_1 and a target Y is very promising as this suggests the predictor may be useful in a model. Now consider another predictor X_2 which has a strong, direct relationship with X_1. In this case, X_2 may not be useful in a model which already includes X_1 as the predictors are redundant (to some degree). A key to variable selection is to add predictors which exhibit a strong relationship with the target but not among themselves.

The regression and classification models that we will cover assume a linear relationship between predictors and the target. Recall that the correlation coefficient measures a linear relationship, and this is appropriate for such models. There are several machine learning models which do not make a linearity assumption, such as random forests and gradient boosted trees. These approaches offer flexibility in using predictors with complex relationships. This also means that while two variables might not have a strong linear relationship, they might have a more complex relationship that can be explained by more flexible models. This highlights the notion of using the right statistic for

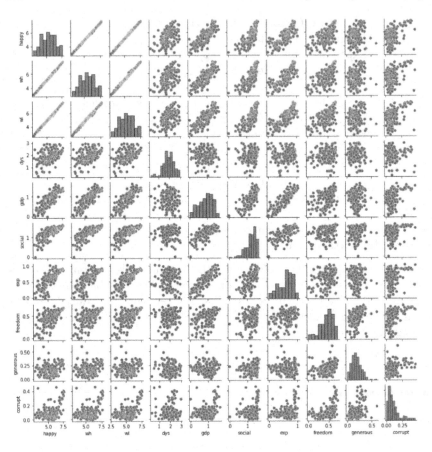

FIGURE 11.10: World Happiness Pairplot

the job, and understanding the limits of a statistic's usefulness. Let's consider two examples:

1. The correlation between a predictor X and target Y cannot tell us whether X caused Y. This is particularly true in observational data, which is not collected under the control of a researcher. A positive correlation indicates that an increase in X is associated with an increase in Y, but we cannot confirm a causative effect. For example, in a study on whether coffee X lowers blood pressure Y for a group of individuals, additional variables like exercise E or stress S may be related to both X and Y and play a causative role. These additional variables are called *lurking variables*, and they can have a *confounding* effect in the study.

2. The mean is not a robust measure of central tendency. For datasets containing large outliers, the mean will be highly influenced by these values and will fail to represent the center of the data. As an alternative, the median should be used in place of the mean when the data is at risk of containing outliers. Examples include financial transactions in real estate, where the sale of an outlier home can distort the overall pricing of a neighborhood.

11.4   Graphical Excellence

In the earlier sections, we used Python modules to produce graphics that seemed helpful in exploring variables. Let's spend some time on what makes for good graphics, and what should be avoided. Many of the points made in this section can be found in the excellent book [35]. Graphics introduce overhead, as they may include elements such as curves, a coordinate system, and labels which must be understood by the viewer. For small datasets, a graphic may not even be necessary; it may be preferable to show the values in a table.

When we decide to produce a graph, it is essential to tell the truth about the data. This may sound obvious, but there are many ways to lie about data, both wittingly and unwittingly. In the first category, visual effects might be intentionally used to mislead a viewer. In the second category, we might use technology that confuses the viewer. In either case, the graphics creator is responsible for accurate, clear reporting of the data. This means applying the appropriate ethics and skill. If a certain package cannot produce the correct graphic, then another package or method should be applied.

Next, we will look at several small examples of plotting data to understand what can go wrong, and how to improve the depiction of the data. Pie charts are very popular for displaying data, particularly in the financial community where they might display the percentage of wealth invested in different assets like stocks and bonds (the breakdown is called an *asset allocation*). Figure 11.11 shows an example of an asset allocation. See if you can guess the data. You might reason that the larger wedge is greater than 50% and less than 75%, and you would be right. It might be hard, however, to arrive at the correct answer: the larger wedge is 65%, while the smaller wedge is 35%. Why make the viewer work so hard and risk arriving at the incorrect answer? The chart could be labeled for greater clarity, but with this little data, it would be better to make a table:

| Stocks | 65% |
| Bonds | 35% |

FIGURE 11.11: Asset Allocation

For the next example, consider plotting the ordered points (1,5), (2,8), (3,10), and (5,12). Suppose we run this code to produce a line plot:

```
xval = [1, 2, 3, 5]    # list of x-coordinates
yval = [5, 8, 10, 12]  # list of y-coordinates

sns.lineplot(x=xval, y=yval)
```

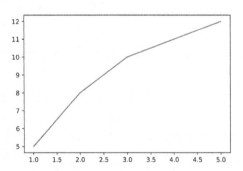

FIGURE 11.12: Misleading Line Plot

While it is simple to produce this graph shown in Figure 11.12, it is highly misleading. It appears that there is continuous data for x-values between 1 and 5, and correspondingly for y-values between 5 and 12. For example, it appears there is a point (4, 11). In fact, there is no such point, but the graph interpolates this value along with many others.

We would do better with a scatterplot, which shows the actual data points, or even a small table.

We also need to take care when using **seaborn** to create bar charts. Here is an example:

```
xval = ['a', 'b', 'c', 'd']
yval = [ 5, 8, 10, 12]
sns.barplot(x=xval,y=yval)
```

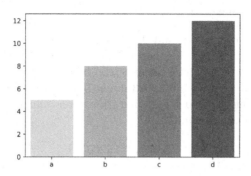

FIGURE 11.13: Misleading Bar Plot

The default behavior is to assign a different color to each data point as shown in Figure 11.13. This effectively creates a *fake data dimension*. We can do better by using one color for all of the bars:

```
sns.barplot(x=xval,y=yval, color='grey')
```

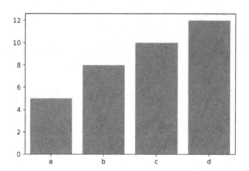

FIGURE 11.14: Better Bar Plot

The plot in Figure 11.14 better conveys the data. To be sure, these points might be better reported in a table, but this small example serves to illustrate the following recommendations:

• Avoid fake dimensions

• Select the right plot for the data

• Question the output of statistical software. Default settings are not always appropriate.

It is important to remember that different viewers may perceive the same graph differently. Color is sometimes used as a variable in graphs, where in particular a color map may suggest severity. Some viewers may have certain kinds of color blindness, and the colors may be misleading. For others, they might struggle to understand the interpretation.

Visuals can act like a double-edged sword: points of interest can garner attention and lead to interest, and they can also mislead. In the next example, we plot revenue over time. The viewer will follow the trend and see that the rightmost value fell. There is a bit of a trap here: the 2023 value is based on the first half of the year. In fact, 2023 might be the best year. It is important to clarify this point to avoid misleading the viewer; annotation can be added to the plot to call out the half year.

```
year = ['2020', '2021', '2022', '1H2023']
revenue = [ 5, 8, 10, 7]
sns.barplot(x=year, y=revenue, color='grey')
```

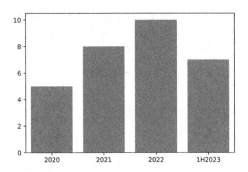

FIGURE 11.15: Bar Plot with Half Year

Adding layers to a plot should only be done if they are helpful in understanding the data. Consider the histogram in Figure 11.16 with an overplotted set of gridlines. The gridlines are so dark that they dominate the plot.
Lighter gridlines as shown in Figure 11.17 allow the data to come forward.

From the examples given above, we see the importance for graphics to be clear and accurate to avoid misleading viewers. As we create graphics to show data, we should challenge them with questions including those listed below. By honestly addressing the questions, we can produce better graphics.

- Is this graphic necessary?

- Is this clearly communicating the data?

- Is there anything misleading, missing, or incorrect?

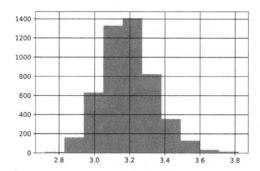

FIGURE 11.16: Histogram with Loud Gridlines

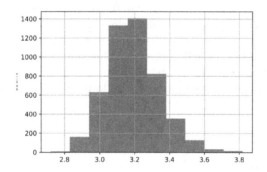

FIGURE 11.17: Histogram with Softer Gridlines

- Is there a better way to label the figure?

- Would there be a better way to show this data?

11.5 Chapter Summary

In this chapter, we discussed techniques and tools for carefully exploring data, which is an essential part of every analysis task. Datasets are often too large for comprehensive manual review, and this is where statistical analysis, aggregation, and filtering can help. Statistics have their limitations, and we discussed for example when the median is a better statistic than the mean. Correlation measures the linear relationship between two variables, and it can help understand if the variables tend to move in similar or opposing directions. We cannot conclude that a strong correlation implies a causative effect, however.

We reviewed a dataset of check deposits, which included the check amount and indicators of whether the check was signed and fraudulent. Using a two-way classification table, we were able to uncover that unsigned checks were more likely to be fraudulent than signed checks. Additionally, by querying the fraudulent checks, we learned that their check amounts were more likely to be multiples of $1000. While this data was synthetic, it illustrated that we needed to be thorough and creative in finding predictors. Different visualizations can also generate insights.

The second exercise was from the World Happiness Report. In addition to happiness scores for over one hundred countries, it included six factors that might explain happiness. We created bivariate plots and computed correlations to understand the relationship between the factors and happiness.

Finally, we discussed important practices for creating useful graphics. Most crucially, graphics must tell the truth about the data. Through both human means and the use of technology, viewers can be misled by graphics. Creators of graphics must be intentional and careful to show what is relevant, leave out what is not, and represent the data clearly, accurately, and efficiently. In doing this, graphics will be additive to understanding.

In the next chapter, we will review machine learning (ML) at a high level. One of the exciting features of ML is giving algorithms the ability to learn from data. This will allow us to formally build predictive models and discover structure in the data.

11.6   Exercises

Exercises with solutions are marked with Ⓢ. Solutions can be found in the `book_materials` folder of the course repo at this location:

https://github.com/PredictioNN/intro_data_science_course/

1. Ⓢ From the Check Fraud exercise, what was the probability of a check being fraudulent given that it was unsigned?

2. The `crosstab()` function below produces a two-way table showing the number of checks that are fraudulent or not, given that the address is valid or not. Based on the counts, what is the probability that a check is fraudulent given that the address is invalid?

```
pd.crosstab(index=df['invalid_address'],
            columns=df[is_fraud])
```

is_fraud	0	1
invalid_address		
0	2	99
1	2050	1

3. Ⓢ For each of the given scenarios, indicate whether the mean, median, or mode would be the most appropriate measure of central tendency:

   a) You would like to compute the central tendency of a set of home sales in a certain neighborhood. The homes that sold consist of 20 nearly identical ranch-style 3-bedroom houses.

   b) You would like to compute the central tendency of a set of home sales in a certain neighborhood. The homes that sold consist of 20 nearly identical ranch-style 3-bedroom houses and one seaside mansion.

   c) A dataset has a categorical variable with some missing values. You wish to use a measure of central tendency to impute the missing values.

4. You hypothesize that stray marks on a bank check make it more likely to be fraudulent. What would be a useful way to explore this hypothesis?

5. True or False: You would like to display four numbers to a viewer. It is generally preferable to show such a small set of data in a table rather than a graph.

6. Ⓢ True or False: The primary objective of graphics should be to show the viewer something visually impressive.

7. You produce a 3-dimensional bar chart where the depth of the chart is not related to the data. Which principle does this violate?

   a) Graphics must be beautiful
   b) Avoid fake dimensions
   c) Question the output of statistical software
   d) Do not show incomplete observations

8. You would like to show the relationship between two variables. Which plot would be most suitable?

   a) Bar plot
   b) Candlestick plot
   c) Pie chart
   d) Scatter plot

9. True or False: A scatterplot indicates a weak correlation between a predictor and target. This suggests that the predictor will not be useful in any machine learning model.

10. Ⓢ You produce a bar plot where each month of the year is represented with a different color. Is this a good practice? Explain why or why not.

11. Ⓢ From the World Happiness dataset, use `seaborn` to produce a scatterplot with the `Generosity` variable on the x-axis and the `Happiness score` on the y-axis.

12. From the World Happiness dataset, you will remove outliers and then compute the correlation between the `Generosity` variable and the `Happiness score`. For this exercise, an observation should be flagged as an outlier if the `Generosity` value is more than $1.5 \times IQR$ below $Q1$ or more than $1.5 \times IQR$ above Q3.

12

An Overview of Machine Learning

In this chapter, we will review the branches of machine learning and the archetypal problems that they solve. Each approach provides a way to learn from data and use it for one or more of these purposes: making a prediction, making a decision, discovering patterns, or simplifying structure. This is fundamentally different from telling the computer how to solve a problem through a set of instructions. Additionally, we will learn about the important elements that arise in solving problems with machine learning. Here are the branches we will explore:

- Supervised Learning

- Unsupervised Learning

- Semi-supervised Learning

- Reinforcement Learning

We will begin with a simple approach to decision making that does not learn from data and does not require an algorithm. This will surface some important concepts, but it will leave us short of solving the problem. It will help us understand why some data problems are too complex to solve using intuition and human number-crunching. We will need tools from machine learning and the help of computers.

12.1   A Simple Tool for Decision Making

There is a good chance we have all used a *pros and cons list* to help make an important decision. The decision might have been whether to select a particular college major, a course, a job, or a place to live. One feature these decisions all have in common is that they are binary: a path is either taken or not. It isn't possible to do half a major in music, and for most of us, spending half the year at a beach bungalow in Santa Barbara is not an option. In making the decision, our goal is to maximize some outcome, like life satisfaction or a long-term career. We will use the decision of where to live as an example, but

I encourage thinking about a decision you personally made using a pros and cons list.

To create the list, we make a *pros* column to contain the advantages of taking the path, and a *cons* column for disadvantages. Then we think about different attributes that we believe will be relevant, what those attributes are like on that path, and how we would perceive them. We may decide that our next home should maximize our overall life satisfaction. Perhaps moderate, sunny weather is important, as well as being near family, having a lot of work opportunities, and the presence of a great food scene. This gives four attributes that seem most important. Are there others that could increase our satisfaction or decrease our satisfaction? This is important, because missing predictors can lead us to the incorrect decision. It turns out that the summer internship last summer included a heinous commute, and so we add commute time to the attribute list.

Let's pause for a minute and answer a question: are we using data in making this list? Well, we might not be directly collecting a massive dataset, but we may research the location's weather, traffic, companies where we can work, and restaurants. We might visit Yelp or Google Maps to collect this information. The other way we are using data is from our mental database of life experiences. We recalled the summer commute, and perhaps what it was like having long winters as a kid. However, we encounter some complications. First, our memory might not be so accurate. The average commute might have been 40 minutes, but we might recall it being 60 minutes. Second, we might respond differently to the attributes in the future than we did in the past. A snowy day as a kid may have involved snowball fights and snow angels, while snow as a recently graduated college student or adult might mean shoveling a driveway or canceling a get-together. In short, we will need to use possibly faulty memories and predictions to measure our satisfaction. Sounds hard, right?

Moving on, we decide that we have all of the important attributes and how we feel about them. It turns out that our location under consideration has the weather, food scene, traffic patterns, and job opportunities that we would like, but it is far from family. Let's construct the list:

Pros
1. Moderate, sunny weather
2. Light traffic and short commute times
3. Many job opportunities
4. Great selection of excellent restaurants

Cons
1. Far away from family

Once again, we remind ourselves to be sure that we are not missing any important attributes. Next, we need to think about how to reach a conclusion, which means somehow aggregating this data into a Go/No Go decision. This is the hard part. Do we count the number of pros and subtract the number of cons? This would assume that each attribute is equally important. Do we say that family is ten times more important than each of the other attributes? Or is the number twenty? One hundred? It becomes apparent that some kind of weighting might play a role, even if each item carries an equal weight. At this point, some people change the weighting to make the decision for them. They increase one of the weights to the point where the decision is obvious. This might be okay so long as the correct decision is reached.

Let's recap and reflect on what we have learned. In making a decision, we collected the relevant attributes. We needed to gather all of them, or risk making the wrong decision. We might have gathered some data and done some research. We likely reflected on past experience, but this was troublesome as our memories can fool us. Additionally, the thought experiment required us to predict how we would perceive somewhat related experiences in a different environment. For experiences very far back in the past, they might not have even been useful. Recency of data is valuable, as the world might have changed from long ago. Lastly, after listing each attribute under the pros and cons columns, we faced the decision of how to weigh them.

I hope you can appreciate how hard this task is. At the same time, I hope you can start to see how machine learning might help. Now, machine learning is not going to tell us what we should feel, but it might help us understand our feelings, and act in a way that is consistent with our goals. How can it do this? We still need to collect data, and we should keep only the data that is relevant. We still need to create the predictors. However, machine learning can be used to determine the optimal weights for any number of predictors. This will require data, a mathematical objective function, an algorithm, and a way for the algorithm to learn from the data. Depending on the algorithm, learning will differ, but in general it follows this flow in the supervised learning case we will discuss next.

1. Start with some weights, which are perhaps randomly assigned

2. Using the weights and predictors, combine them to make a prediction. Each predictor might be multiplied by its weight, and then the terms can be summed.

3. Compare each prediction to the correct answer. For example, differences might be computed. This returns errors.

4. Change the weights to reduce a measure of the total error. The measure might be the sum of squared errors, for example.

5. Repeat 2-4 until we run out of permitted iterations or the improvement is marginal

Now, let us study supervised learning in more detail. This is a classical machine learning approach that has wide applicability.

12.2   Supervised Learning

Supervised learning (SL) uses n data pairs $\{(\mathbf{x}_1, y_1), (\mathbf{x}_2, y_2), \ldots, (\mathbf{x}_n, y_n)\}$ to learn a function that can predict a target y given $\mathbf{x}$. We denote the labeled dataset as $\{(\mathbf{x}_i, y_i)\}_{i=1}^{n}$. It is common in classification problems (which we discuss shortly) to refer to target values as *labels* or *ground truth values*. For a given observation i, the data pair $(\mathbf{x}_i, y_i)$ consists of a d-dimensional vector of predictors $\mathbf{x}_i$ associated with the label y_i. When each observation has a label, we say that the dataset is *labeled*.

Taking a step back, we might ask if it is possible to avoid learning a function and take a simpler approach: why not use this data as a lookup table? That is, if we were given a new $\mathbf{x}^*$, could we look up the data pair for the right $\mathbf{x}$, and return its y? There are some problems with this approach. First, the particular value of $\mathbf{x}^*$ might not be in the dataset. Second, if $\mathbf{x}^*$ is in the dataset, the value might be noisy, leading to an erroneous y. Third, for a case with a large set of data pairs with possibly high-dimensional $\mathbf{x}^*$, this could be a massive lookup table. Given these shortcomings, it motivates learning a function to handle the mapping for us. It can learn from the data at hand, interpolate for new values of $\mathbf{x}^*$, and complete much faster as it avoids lookup. What we give up using this approach is that the function is an approximation, and we must select a metric to minimize the error.

There are two kinds of supervised learning problems, and this depends on whether y is a continuous- or discrete-valued variable. In the continuous case where y takes an uncountable number of values, the problem is a *regression problem*. An example is predicting the price of a house. In the discrete case where y takes a countable number of values, the problem is a *classification problem*. A common task is to predict an outcome with two possible values, or *classes*, such as:

- The patient recovers (the *positive class*) or not (the *negative class*)

- A bank deposit is fraudulent (positive class) or not (negative class)

- A stock price rises over the next trading day (positive class) or not (negative class)

These examples are called *binary classification problems*. The positive class is assigned to the event of interest; it makes no judgment on good versus bad. Programmatically, the positive class is coded as 1, and the negative class as 0.

In the case where the outcome has more than two possible values, it is called a *multiclass classification problem*. Examples might include predicting the fastest runner in a marathon, or the next number rolled on a six-sided die. Coding proceeds by numbering each class from zero to the total classes minus one. For the six-sided die problem, the classes would be numbered $0, 1, \ldots, 5$.

Supervised learning is a very powerful method for predicting outcomes, but it has some strong requirements. First, labeled data is required, and it must be accurate. Depending on the problem and the explanatory power of the predictors, the necessary dataset may be massive. Additionally, the labeled data needs to be relevant. Let's think about why this can be challenging. Consider a new product launch, where the task is to predict the number of units sold in the first month. Since the product is new, there is no data. Now consider a product which has been on the market for ten years. There is an abundance of labeled data. Then the world is gripped by a pandemic, and demand surges. The historical database is now irrelevant. For these reasons, supervised learning might not work.

12.3   Unsupervised Learning

Unsupervised learning problems aim to understand the structure of data that is not labeled. In this case, we have collected a set of observations $\{\mathbf{x}_1, \mathbf{x}_2, \ldots, \mathbf{x}_n\}$ where $\mathbf{x}_i$ is again a d-dimensional vector of predictors. As shorthand, we will denote the dataset as $\{\mathbf{x}_i\}_{i=1}^n$. We can ask questions of this data such as:

- Which of these data points are outliers? This can be used for outlier detection to flag suspicious activity, for example.

- If we wanted to group the data points according to some distance metric, how should they best be grouped? This can help discover substructure to understand different personas in the data, for example.

- For a new data point, which group provides the best fit? For example, in an ideal case, we might find that bank transactions form two groups: fraudulent transactions and non-fraudulent transactions. The new data point, based on the values of the predictors, might clearly belong to the non-fraudulent group.

Unsupervised learning can often be an iterative procedure, as the algorithms require parameters such as the number of groups which often aren't known in advance. In some cases, the grouping can be used as a new variable in analysis. The most popular algorithm for unsupervised learning is *k-means*, and we will discuss this in detail later.

12.4 Semi-supervised Learning

Semi-supervised learning (SSL) uses a combination of supervised and unsupervised learning. In some cases, there may be a small set of labeled data and a much larger set of unlabeled data. We will review one approach to SSL. There are several other techniques, and the interested reader can consult [36].

Consider a binary classification task. To get the process started, the small set of labeled data may be used to train a supervised learning model. Next, the unlabeled data can be run through the model, or *scored*. When the unlabeled data is scored, each data point is given probabilities of belonging to the positive class and the negative class (more on this later). For the data points with a sufficiently high probability, they can be assigned the corresponding label. Table 12.1 shows an example of three data points and their probabilities of originating from the positive and negative classes. Assume a required threshold of 0.90 for making a label assignment. In this case, data point 1 would be assigned to the positive class, as its probability of 0.91 exceeds 0.90. The second data point would be assigned to the negative class as its probability of 0.96 exceeds 0.90. The third data point would not be assigned, as neither of its probabilities exceed the threshold.

TABLE 12.1: Labeling data points based on probabilities.

Data Point	P(neg class)	P(pos class)	Action
1	0.09	0.91	Assign to positive class
2	0.96	0.04	Assign to negative class
3	0.60	0.40	Low confidence, do not assign

If possible, the data points to be assigned labels should be reviewed for quality control. After the newly labeled points are added to the training set, the model can be refit. This process can be repeated to increase the size of the training set. After some amount of iteration, it might not be possible or necessary to continue iterating, and the trained model can be used.

12.5 Reinforcement Learning

In *reinforcement learning* (RL), an agent must decide on the best sequence of actions to take over time to maximize an objective. The agent begins in state s_0, takes action a_1, receives reward r_1, and transitions to the next state s_1. This continues until a stopping condition, generating tuples $(s_t, a_{t+1}, r_{t+1}, s_{t+1})$ which can be collected and used for training the system. For actions leading to good outcomes, the reward will be positive, while bad outcomes will produce negative rewards as a penalty. Some important components to be determined are:

- The *state space* S representing the important variables

- The *action space* A representing the set of actions that the agent can take at each time point

- The *reward function*

- The terminating condition, which ends the sequence

- The *discounting factor* applied to the reward, as immediate positive rewards are more valuable than rewards further into the future

- How the state changes when the agent takes a particular action from each state. This is determined from the *environment*.

- The *optimal policy*, which is a function that returns the best action for each state

To determine the optimal policy, the long-term value of the actions needs to be computed. Thinking long term is essential, because selecting the action that produces the highest next reward (*greedy behavior*) might be suboptimal in the long run. The long-term value is computed as the discounted expected cumulative reward.

As a quick example, I used to love playing volleyball in high school. Fast forward fifteen years, and I found myself playing a pickup game at the local YMCA after not playing since college. The game lasted three hours and I loved it. Compared to all the other sports I played, including basketball and tennis, I ranked this at the top. That was, until I woke up the next morning and could hardly walk for the next two days, let alone do other activities like weightlifting. I realized that the short-term value of this action was very different from its long-term value!

Another important concept in RL is *exploration* versus *exploitation*. For example, imagine deciding the best Chinese restaurant in town by trying each of them once. Perhaps restaurant A is the best restaurant but had an off night, while restaurant B performed on par. Greedy behavior would dictate that the agent should avoid restaurant A and return to restaurant B. This is exploitative behavior, and it will produce a suboptimal policy, as restaurant A is the better restaurant on average.

An exploratory strategy encourages the agent to sample each action to learn the optimal policy. In an ϵ-greedy approach, the agent explores for a small fraction of the time ϵ and takes the highest-valued action for the remaining majority of the time $1 - \epsilon$. For the portion of time when the agent explores, it randomly selects from all available actions. This allows the agent to generally benefit from its learnings, while taking some time to explore actions which may be more valuable.

The reinforcement learning approach has several benefits compared to supervised learning:

- It mimics how we learn from experience

- Labeled data is not required. For some problems, such as the best way to treat patients for a life-threatening condition like sepsis, the right answer is not known. RL can help guide the patient to improvement based on observed rewards.

- It models a sequence of actions for a complex process

- Learning can happen in a dynamic environment

Depending on the problem, there can also be several challenges such as:

- The environment needs to be known, accurately modeled, or simulated. In some cases, it is not possible to test and learn, such as in medicine where patients cannot be randomly assigned different actions.

- There can be a time delay between taking the action and observing a change in state. As several actions might be taken before seeing the change, this makes it difficult to attribute the change to a particular action.

- For some problems, the state space S, the action space A, or both might be continuous. This increases the computational cost and necessitates different algorithmic approaches.

Examples of RL problems include:

- A doctor providing a series of medication doses to a patient to bring a lab value into a reference range. The state space is the lab value measurement, the action space might be whether to increase the dose, decrease the dose, or keep it unchanged, and a reward function might give positive values to actions that move the lab value into the range or closer to it. For actions moving the lab value away from the range, a penalty would be assigned. The terminating condition would occur if the patient faces mortality.

- A portfolio manager selecting fractions of wealth to invest in a universe of stocks over time to maximize return. The state space is the fraction invested in each stock and the action space might be whether to increase the allocation, decrease the allocation, or keep the allocation unchanged for each stock. To simplify the action space, the investor might only be allowed to change each allocation by a fixed percentage at each time step. The reward can be the cumulative return, which is the portfolio's percentage change in value since the start. The terminating condition might be the earlier of the end of the investment horizon and the portfolio value going to zero.

- Navigating a drone to a safe landing zone while avoiding obstacles. The state space is the location of the drone in three spatial dimensions (perhaps surrounded by a bounding box) and the action space is the direction taken with a remote controller. The reward might be the negative straight-line distance to the landing zone (i.e., moving closer to the landing zone is better). The epoch can terminate if the drone strikes an obstacle.

This sketch should provide an idea of the reinforcement learning approach, and how it is different from supervised and unsupervised learning. This is a very active research area that has led to fruitful results. A deeper discussion of RL is outside the scope of this introductory text on data science. For learning more, please see [37].

12.6 Generalization

We will focus most of our modeling attention on the supervised learning approach. A critical assumption in SL is that the patterns learned from the training set will generalize to additional data. The additional data might be a separate segment of data that has already been collected, or data yet to be collected.

One of the greatest fears of a data scientist is that a model is trained on some data, it fits the data well, it is deployed, and it fails to work as well in production. This would mean that the model learned relationships between the predictors and target from the training set, but those relationships were not present in production data. This can happen, and when it does, it is a major disappointment.

How can we avoid this lack of generalization? Unfortunately, it can't be eliminated with certainty, but there are steps a data scientist can take to improve the chances of generalization. First, it is important to distinguish between two cases:

1. The model fails out of the gate. This is often preventable.

2. The model works in production for some time and later fails

In the first case, the problem is often caused by *overfitting*. As an example, imagine cramming for an exam by taking practice tests. If we learn the practice exams perfectly, we can score perfectly on the real exam only if it matches the practice questions. Suppose there are some quirky practice questions that don't deeply test the core concepts. We might learn those and have a false sense of confidence that we know the material. It is unlikely, however, that these questions will be recycled on the real exam.

Model overfitting can have the same effect as cramming the quirky questions. While the model is learning some patterns, it is also memorizing idiosyncratic effects that are unlikely to appear again. This can fool us into overestimating the model's ability. As an example, consider Figure 12.1. There is a data point at $x = 3.5$, $y = 9$ that stands apart from the rest of the data which exhibits a clear linear relationship. A good model will fit the linear trend and ignore the outlier. It will generalize well to new data, assuming that it continues to follow the relationship $y = 2x$. An overfitting model will attempt to fit the outlier, which will take it off course from a linear relationship. As it is unlikely that future data will follow the outlier, the more complex model will not fit as well as the linear model.

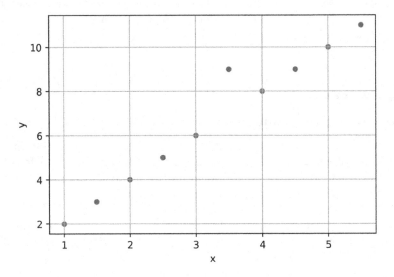

FIGURE 12.1: Data at Risk of Overfitting

A powerful way to guard against overfitting is by splitting the dataset into pieces and using them strategically to fit and estimate model performance. There are different methods with varying degrees of complexity, so let us start with a relatively simple training/validation/test scheme. In this setup, the dataset is broken into three parts:

- The *training set* is used for training the parameters, or weights, in the model

- The *validation set* is used for tuning *hyperparameters* in a model. A hyperparameter is an external configuration variable which is used to control the learning process. Several models use hyperparameters, and their optimal values are not known in advance. As an example, some models build decision trees, and the number of trees to build is a hyperparameter. We will review hyperparameter tuning shortly.

- The *test set* (or *holdout set*) is used for evaluating the model performance after all of the modeling decisions have been made. That is, the model, the parameters, and the hyperparameters all need to be decided at this point. The test set is then useful for giving an independent assessment of how the model might perform on new data.

The test set is critical as the model performance on the training set will overestimate its ability. Stated differently, we can expect that the performance on the test set will be worse than the performance on the training set, and it will be more realistic. If the performance is measured on the test set and then changes are made to the model, then another test set will be needed for a true measure of performance. For this reason, multiple test sets are sometimes formed, leading to the training/validation/test1/test2 scheme.

The process of breaking the data into pieces is called *data splitting*, and it is an essential step in the machine learning process. This should take place after the data is cleaned and processed, but before modeling. The separate pieces should be saved into separate named files such as train.csv, validation.csv, test.csv.

Care needs to be taken when splitting the data. Since the pieces of data need to be independent from one another, we will need to think about what each row represents. For example, suppose each row represents a patient measurement at a snapshot in time. This allows for the possibility that one or more patients may have multiple records in the dataset. If we split the data by randomly selecting rows, this can introduce an issue called *data leakage* where one or more patients can have portions of their data in both the training set and test set. This is a problem because the model will learn on the training set and will be evaluated on a test set that is no longer independent of the training data. Overstated performance can result from data leakage. The issue can be circumvented by getting the unique list of patient identifiers and randomly assigning fractions of them to each data subset.

As an example, we might decide to allocate 60% of the data to the training set, 20% to the validation set, and 20% to the test set.[1] Operationally, we would build the list of unique patient identifiers (a one-liner of Python code), randomly assign patient identifiers to the three sets in the given proportions, and move them with their associated data to create the datasets. In this way, each patient's records belongs fully to either the training set, validation set, or test set without risk of leakage.

Finally, let's return to the point that even if a model is carefully trained on the training set and performs well on an independent test set, it can underwhelm later. This is a case where the model generalized well to unseen data at first, but its performance degraded later. Data scientists should expect this to happen, actually. The value of data decreases over time, and thus, the value of models will decrease over time. Patterns will change for many reasons, including:

- A pandemic

---

[1] It is common to use a majority of the data for training and 60% or 70% are common values, but not the final word.

- Structural changes in the economy

- Geopolitical factors

- Changes in consumer preferences

- Business strategy changes

As a quick example of a business strategy change, consider a model used by a commercial bank to decide if someone should be approved for a mortgage. One of the strongest predictors of default, or non-payment in part or in full, is a borrower's credit. Suppose that at the time of model development, the bank targeted borrowers with credit scores above 750. When the training set was collected, most of the borrowers met this credit score requirement, and the model was able to accurately predict default for this subpopulation. One year later, the bank was struggling to find enough of these stronger borrowers, and they lowered their credit standards to a 700-score minimum. Recall that applicants with scores in the 700-750 range were not a priority at the time of development; in fact, they represented a very small fraction of the training set. Under the new strategy, many of these under-represented borrowers were given loans, and they defaulted at a much higher rate than predicted. This is not surprising, because the distribution of credit scores in the training set was different from the borrowers under the new strategy. To combat these effects, it will be important to put a model monitoring strategy in place, and update the model regularly with fresher data.

12.7   Loss Functions

Recall that supervised learning uses training data $\{(\mathbf{x}_i, y_i)\}_{i=1}^{n}$ to learn a function that can predict the target variable. The predictions $\{\hat{y}_1, \hat{y}_2, \ldots, \hat{y}_n\}$ of the true values $\{y_1, y_2, \ldots, y_n\}$ may differ, producing errors $\{\epsilon_1, \epsilon_2, \ldots, \epsilon_n\}$.

We need a useful way to measure the cost of making each error, and to combine these costs. Once we have established this, we can work toward improving the model by changing its parameter values. The *loss function* will be used for this purpose. It will depend on the actual value y_i and its prediction $\hat{y}_i$ for a given observation i. The loss function will be denoted $L(y, \hat{y})$, and we must select an appropriate form. As $\hat{y}$ is produced by a model or functional form f which depends on predictors $\mathbf{x}_i$ and parameters $\boldsymbol{\theta}$, we can write the loss function as $L(y, f(\mathbf{x}_i, \boldsymbol{\theta}))$ to show explicit dependence on the parameters to be optimized.

As the loss function measures the cost of an error from a single observation, and y is treated as a random variable following a distribution, we will want to compute its expected value. This quantity is called the *expected loss* or *risk*. In practice, the expected loss can be estimated empirically by averaging the loss over all of the training observations. The objective then becomes minimizing the quantity R_{emp} defined as:

$$R_{emp}(\boldsymbol{\theta}) = \frac{1}{n} \sum_{i=1}^{n} L(y_i, f(\mathbf{x}_i, \boldsymbol{\theta}))$$

This process is called *empirical risk minimization*. The fact that this operation yields the appropriate estimator is an important result in machine learning. Details can be found in [38].

There is flexibility in selecting a loss function, but some attributes are important:

- Errors cannot be measured as signed distances, as summing them would allow for cancellation

- It is preferable for the function to be continuously differentiable. This is because we will compute its gradient on the way to minimizing expected loss with respect to the function parameters. At points where a function is not differentiable, the gradient will not be defined.

Beyond this, the choice of loss function can depend on whether the SL task is regression or classification. The *squared error*, or $L2$ loss, is a popular choice for the regression task. Since the expected loss needs to be measured, the squared error can be averaged over the training set to produce the *mean squared error* (MSE) as follows:

$$MSE = \frac{1}{n} \sum_{i=1}^{n} (y_i - \hat{y}_i)^2$$

Given that the errors $y_i - \hat{y}_i$ are squared, their contributions are relatively small between -1 and 1, but they grow quadratically for larger values. This means that outliers will have a large influence on MSE (it is not a robust metric). Mathematically, MSE is continuously differentiable, making it straightforward to minimize with respect to parameters.

Next, let's look at a small example of calculating MSE with code.

```
import numpy as np

y_actual = np.array([1.1, 0.5, -1.2])
y_hat = np.array([1.2, 0.3, -1.3])

def mse(y, yhat, logging=True):

    # compute errors
```

```
error = y - yhat

# squared errors
sq_error = error ** 2

mse = sq_error.mean()

if logging:
    print('error:', error)
    print('sq_error:', sq_error)
    print('mse:', mse)
return mse
```

First, we define a numpy array of actual values and corresponding predictions. We define the mse() function, which takes the actuals, predictions, and an optional parameter named *logging* to allow for step-by-step output; logging is given a default value of True, which means the logging block will run unless a value of False is passed as a third argument. Here is the function call and output:

```
mse(y_actual, y_hat) # function call

error: [-0.1  0.2  0.1]
sq_error: [0.01 0.04 0.01]
mse: 0.02
0.02
```

Notice that the MSE of 0.02 is smaller than each of the errors. This is because this loss function is squaring errors which are fractions.

The *absolute error*, or *L*1 loss, is another common choice in the regression task. We can average the absolute errors over the training set for an estimate of the expected loss. This metric is called *mean absolute error* (MAE), and it is defined as follows:

$$MAE = \frac{1}{n} \sum_{i=1}^{n} |y_i - \hat{y}_i|$$

Since the absolute value is computed on the errors $y_i - \hat{y}_i$, their contributions grow linearly. Outliers will have less influence on MAE when compared to MSE. MAE is more challenging to minimize, however, as the function is piecewise linear and not differentiable when the error is zero. Next, let's look at code to calculate MAE. We will use the same data for comparability.

```
def mae(y, yhat, logging=True):
    error = y - yhat
    abs_error = np.abs(error)
```

```
mae = abs_error.mean()
if logging:
    print('error:', error)
    print('abs_error:', abs_error)
    print('mae:', mae)
return mae
```

Aside from variable names, the only meaningful difference is the line that computes the loss. Now let's see the output:

```
mae(y_actual, y_hat)
```

```
error: [-0.1  0.2  0.1]
abs_error: [0.1 0.2 0.1]
mae: 0.13333333333333333
0.13333333333333333
```

For this case, as the errors are fractions, MAE is larger than MSE. Let us now compare squared error loss to absolute error loss across a range of errors. Figure 12.2 represents squared error with a solid curve and absolute error with a dotted curve. This confirms what was explained earlier: for fractional errors, absolute loss will be greater. Outside of this range, squared loss will dominate.

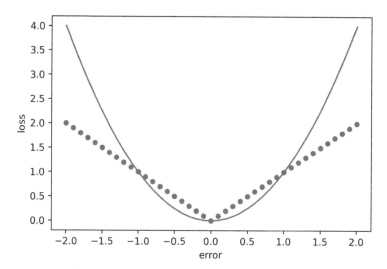

FIGURE 12.2: Loss Functions: Squared Error versus Absolute Error

Next, let's turn to the binary classification task. Let $\hat{p}_i$ and $1 - \hat{p}_i$ denote the predicted probability that observation i has positive and negative label, respectively. The true label for this observation is y_i. The *cross-entropy loss* is defined as:

$$H(\hat{p}_i, y_i) = -(y_i log(\hat{p}_i) + (1 - y_i) log(1 - \hat{p}_i))$$

The expected loss can be computed by averaging the cross-entropy losses over each observation in the training set. For the case where y_i=1, the second term vanishes and we are left with $-log(\hat{p}_i)$. For the case where y_i=0, the first term vanishes and we are left with $-log(1 - \hat{p}_i)$. Therefore, the quantity reduces to the negative log of the predicted probability of the actual label.

To gain more intuition with cross entropy, let's think about when it attains its minimum value of zero. There are two cases where this happens:

- the observation has a positive label and the predicted probability of the positive label is 1: $y_i = 1$, $\hat{p}_i = 1$

- the observation has a negative label and the predicted probability of the negative label is 1: $y_i = 0$, $1 - \hat{p}_i = 1$

When there is disagreement between the actual label and the predicted probability, there will be a loss. Figure 12.3 shows the cross entropy when $y_i = 1$ for different values of $\hat{p}_i$. When $\hat{p}_i = 1$, the loss is zero. As the probability decreases, the model is less confident that the label is 1 and the loss increases asymptotically.

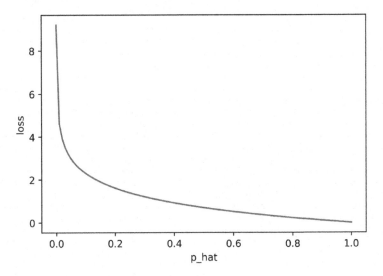

FIGURE 12.3: Cross-Entropy Loss when $y_i = 1$

We briefly reviewed some common loss functions in this section. When fitting regression and classification models in Python (for example with `sklearn`), it will not be necessary to explicitly code the loss functions. It is useful, however, to understand how they work.

12.8   Hyperparameter Tuning

As mentioned earlier, some models use hyperparameters, and their best values are not known in advance. Some hyperparameters are integer-valued, such as the number of decision trees to grow, while others are real-valued, such as the penalty for using large parameter values. A common practice for learning the optimal hyperparameters is *grid search*. This is an exhaustive search method that works like this:

1. Provide a list of values for each hyperparameter

2. For each combination of hyperparameter values, fit the model

3. Evaluate model performance for each combination of hyperparameter values

4. Determine which hyperparameter combination produced the best model performance. These are the optimal hyperparameters.

The benefit of grid search is that it is comprehensive, as it tries all combinations of listed values. However, there are some shortcomings:

- For real-valued hyperparameters, it won't be possible to try all values. A common approach is to try different orders of magnitude (e.g., 0.01, 0.1, 1, 10, 100)

- Since the Cartesian product of hyperparameter values must be evaluated, this method is expensive. For example, if

```
H1: [0.01, 0.1, 1, 10, 100]
H2: [1, 2, 3]
```

grid search will proceed by trying each combination:

```
(0.01, 1), (0.1, 1), ..., (100, 3)
```

There are 5 x 3 = 15 combinations to try, which results in fitting and evaluating 15 models.

- The method will continue to evaluate areas of the search space that don't look promising. This is because it doesn't use information about the performance results.

Alternatives to grid search include Bayesian methods and random search. Bayesian methods focus attention on promising areas of the search space, while avoiding areas that don't result in strong performance.

Earlier, we reviewed a model evaluation scheme where we split the data into train/validation/test. We discussed training the data on the training set, tuning on the validation set, and evaluating the model on the test set after all decisions were made.

A more thorough way to evaluate model performance is *k-fold cross validation*. The idea is to break the data into k equal pieces and take turns evaluating model performance on each of them. Figure 12.4 illustrates the procedure for a typical value of $k = 5$. For each iteration, 4/5 of the data is randomly assigned to the training set and 1/5 is randomly assigned to the validation set. The portion of data in the validation set is different for each iteration. The model is trained on the training data, and a metric such as MSE is computed on the validation data. This results in 5 measurements of MSE (one for each validation set), and they are averaged for an overall MSE estimate.

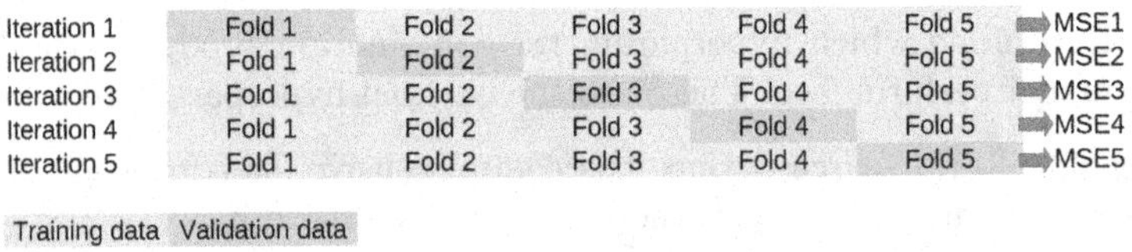

Iteration 1	Fold 1	Fold 2	Fold 3	Fold 4	Fold 5	➡MSE1
Iteration 2	Fold 1	Fold 2	Fold 3	Fold 4	Fold 5	➡MSE2
Iteration 3	Fold 1	Fold 2	Fold 3	Fold 4	Fold 5	➡MSE3
Iteration 4	Fold 1	Fold 2	Fold 3	Fold 4	Fold 5	➡MSE4
Iteration 5	Fold 1	Fold 2	Fold 3	Fold 4	Fold 5	➡MSE5

Training data   Validation data

FIGURE 12.4: K-Fold Cross Validation Illustration

K-fold cross validation is a best practice for model evaluation and hyperparameter tuning, but it can be expensive. From our earlier example with 15 combinations of hyperparameters, consider running 10-fold cross validation. Each combination will require fitting and evaluating 10 models, for a total of 150 models. For a massive dataset, it might be sufficient to use $k = 5$ or even $k = 3$.

The `sklearn` module provides functionality for grid search through the `GridSearchCV` function. One of the important inputs is the grid containing hyperparameter lists. Here is a small example:

```
from sklearn.model_selection import GridSearchCV
from sklearn.linear_model import LogisticRegression

grid = {'C':[0.01, 0.1, 1, 10], 'penalty':['l1','l2']}
model = LogisticRegression()
model_cv = GridSearchCV(model,grid,cv=10)
model_cv.fit(xtrain, ytrain)
```

The grid is a dictionary containing two hyperparameters: 'C' and 'penalty.' A logistic regression model[2] is instantiated, and the model, the grid, and a splitting strategy parameter 'cv' are passed to the `GridSearchCV` object. The

---

[2]We will study logistic regression in detail later.

'cv' parameter can be used to specify the number of folds to use in cross validation (10 in this case). The `fit()` function will train the model using each hyperparameter combination, and a performance metric is calculated. The final model will use the optimal hyperparameters which produced the best metric.

12.9 Metrics

We examined loss functions, which are designed to measure errors and optimize model parameters. For the purpose of understanding how well a model performs, not all loss functions are easily understood by a human. Cross-entropy loss, for example, is a number that approaches infinity as the predicted probability of the correct label approaches zero. For this reason, we use metrics to measure and report on model performance, and the relevant metrics will depend on the task.

For the regression problem, the MSE and MAE are both reasonable metrics. As MSE represents a sum of squared errors, the square root is often taken, producing the *root mean square error* (RMSE). The RMSE then measures the average error, irrespective of the error directions.

Another useful regression metric is *R-squared* (R^2), and it represents the fraction of variation in the target variable explained by the predictors. R^2 falls in range $[0, 1]$, where 0 indicates that the predictors explain none of the variation (and a useless model), and 1 indicates full explanation. It is difficult to specify a good R^2 in advance as it is problem dependent, but when comparing two models, higher R^2 is better, all else equal.

Increasing the number of parameters in the model can inflate the R^2 without increasing explanatory power. To combat this, the *Adjusted R-squared* can be computed. We will discuss the detailed calculation of R-squared and Adjusted R-squared in the next chapter.

For the binary classification task, each outcome and its associated prediction will take values 0 or 1. Useful metrics will quantify the number and fraction of predictions that are correct (these are cases where the actual label is equal to the predicted label) and incorrect (cases where the actual label is not equal to the predicted label). The fraction of correct predictions is called the *accuracy*, which is simple to compute but sometimes not the best metric. After introducing additional metrics, we will learn why this is the case.

Let's introduce some notation and classification metrics:

- P: a positive value, which can be used for the actual or predicted label

- N: a negative value, which can be used for the actual or predicted label

- TP *(true positive)*: the positive label is predicted, and this is the true label

- FP *(false positive)*: the positive label is predicted, but actual label is negative

- TN *(true negative)*: the negative label is predicted, and this is the true label

- FN *(false negative)*: the negative label is predicted, but the actual label is positive

To summarize, there are two outcomes that are correct (TP and TN), and two outcomes that are incorrect (FP and FN). Which error is worse – a false positive or a false negative? This depends on the problem. In the case of a COVID-19 test, a false positive would cause unwarranted worry. A false negative would suggest that the person isn't infected, which could lead to spreading the disease and not seeking treatment.

Let's imagine that 100 individuals take a COVID-19 test, and the results are TP=85, FP=5, TN=7, and FN=3. We notice that most individuals tested positive, which could make sense if they don't feel well and this prompts the test. From this information, we can compute quantities such as:

- The number of positive predictions: $TP + FP = 85 + 5 = 90$

- The number of negative predictions $TN + FN = 7 + 3 = 10$

- The number of actual positives: $TP + FN = 85 + 3 = 88$

- The number of actual negatives: $TN + FP = 7 + 5 = 12$

This information can be organized into a two-way table called a *confusion matrix*, which has this form:

predicted	P	N
actual		
P	TP	FN
N	FP	TN

For example, the cell FN represents the number of false negatives. This is where the actual label is P and the predicted label is N. We can fill in the confusion matrix with our data:

predicted	P	N
actual		
P	85	3
N	5	7

In this case, there were 3 false negatives where the actual label was P and the predicted label was N. The diagonal of the table holds the correct counts (85+7), while the errors appear on the off diagonal (3+5). The accuracy can be computed as follows:

$$accuracy = \frac{TP + TN}{TP + FP + TN + FN}$$

For this example, accuracy = (85+7)/(85+7+3+5) = 0.92.

Next, let's discuss *recall* and *precision*. Recall is the fraction of true positives that are predicted as positive. It measures how good the classifier is at detecting actual positives. The trouble is that a model can always predict positive and attain perfect recall. As a counterbalance, we use a second metric.

Precision is the fraction of predicted positives that are correct. It measures the accuracy of positive predictions. The trouble here is that the classifier can make positive predictions only when it is very confident, attaining high precision. However, it would then miss most of the actual positives, resulting in low recall. For a classifier to be strong, it must have high precision and high recall. When it predicts the positive label, it needs to be accurate (precision), and it also needs to identify the positive cases (recall). Precision and recall can be calculated as follows:

$$precision = \frac{TP}{TP + FP}$$

$$recall = \frac{TP}{TP + FN}$$

For our example, precision = 85/(85+5) = 0.944 and recall = 85/(85+3) = 0.966.

The last metric we will discuss here is the *F1 score* (F_1), which is the harmonic mean of precision and recall:

$$F_1 = 2 \times \frac{precision \times recall}{precision + recall}$$

As a single number combining recall and precision, the F1 score is very popular and effective for binary classifier evaluation. Additionally, the harmonic mean is more punitive than the arithmetic mean, which results in a conservative metric. As an example, for a classifier with recall=0.2 and precision=1.0, the arithmetic mean is 0.6, while the harmonic mean is 0.333.

The `sklearn` module provides functions for computing regression and binary classifier metrics. However, it is very valuable to understand what they mean so they can be used and interpreted appropriately.

> **Tip: Why accuracy is not always the right metric**
>
> In a binary classification problem, it is generally important to accurately predict both the positive and negative labels. We can consider the case where one of the labels outnumbers the other, and without loss of generality we will take the positive labels to be rare. In this case, the data will be imbalanced and most observations will have negative labels. A classifier can have high accuracy by predicting all observations to have negative labels. Such behavior will not identify any of the positive labels, which means it will not be a useful classifier. This is why the F1 score can be a more useful metric than accuracy for binary classification.

12.10 Chapter Summary

We began this chapter with a simple tool for making a decision: a pros and cons list. We learned that to make a decision, we needed to surface all of the important predictors and determine their weighting. It is not an easy task to determine the optimal weighting, and this is one place where machine learning can help.

Next, we briefly reviewed the branches of machine learning, starting with supervised learning. This approach pairs predictors with labels, so that a function can be learned for taking new predictor values as input and predicting the target. It is not always easy or even possible to collect labeled data, and this is one motivation for other approaches.

Unsupervised learning can be used to understand the structure of data that is not labeled. Given a set of observations containing predictor values, we can try to identify outliers and groupings of the data points. We can also predict the group membership of a new data point. For the case where we have a small fraction of labeled data and a much larger fraction of unlabeled data, semi-supervised learning can be a useful approach. One recipe is to train on the labeled data and predict the labels of the unlabeled data. Reinforcement learning takes a different approach that follows human learning. An agent takes an action and receives a reward and new state from the environment. Working to maximize long-term reward, the agent learns the value of taking each action from each state.

Next, we focused our attention on supervised learning and one of its biggest risks: failure to generalize to new data. Overfitting can be a major culprit, and we discussed mitigating practices such as evaluating model performance on held-out data.

For measuring prediction error, we defined some common loss functions. To measure the error across the training set, we defined risk as the expected loss; operationally we can average the loss over the data points in the training set.

The optimal parameter values will minimize risk, and optimization techniques can be used to find these values.

We discussed hyperparameter tuning and k-fold cross validation. Models often include hyperparameters which are not known in advance and must be fine-tuned. This can be an expensive step, particularly when the search space is large and grid search is used. K-fold cross validation provides a thorough method for evaluating model performance, as each fold of the training data is held out for validation. The technique can be combined with grid search for hyperparameter tuning, and this is a best practice in machine learning.

Metrics are used to report on model performance. We discussed important metrics for the regression problem, such as RMSE and R-squared, and the binary classification problem, such as recall and precision. We will see these metrics again as we work with regression and classification models.

This chapter was very dense with concepts. In the next chapter, we will look at linear regression modeling in detail. Many of the concepts we just studied will be reinforced through code and examples.

12.11   Exercises

Exercises with solutions are marked with Ⓢ. Solutions can be found in the book_materials folder of the course repo at this location:

https://github.com/PredictioNN/intro_data_science_course/

1. Discuss a pros and cons list that you have made, or a list that you could make. What was the decision you were trying to make? What were the pros, what were the cons, and how did you ultimately combine them to produce an answer?

2. Ⓢ Discuss the major difference between supervised learning and unsupervised learning. What is one reason why unsupervised learning might be used in favor of supervised learning?

3. Given a dataset, how do we categorize whether a supervised learning problem should be a regression problem or a classification problem?

4. Ⓢ For each problem below, indicate whether it is a regression problem (R), classification problem (C), or clustering problem (S).

   a) Summer campers are signing up for club activities. The activity each camper will select will depend on interests and who is teaching the activities. You are asked to predict the probability that a camper signs up for each activity.

b) A teacher wishes to place students into three reading-ability groups based on a reading comprehension exam.

c) An engineer needs to predict the engine horsepower of a prototype sports car based on the size of the engine and the number of cylinders.

5. Ⓢ Define a function that takes 1) the positive class predicted probability and 2) the label as inputs and returns cross-entropy loss. Use base e for the logarithm. Next, call the function, passing 0.5 as the positive class predicted probability and 1 as the label. Show your code and the result.

6. A data scientist has a dataset with 10,000 observations. Only 500 observations are labeled. Outline a method using semi-supervised learning to label the remaining observations.

7. Ⓢ Explain exploration and exploitation. What is the downside to an agent that only exploits and never explores?

8. Ⓢ Explain overfitting. Why is overfitting a major concern for data scientists and users of data science models?

9. What is one way that data scientists can guard against overfitting?

10. Ⓢ Explain the function of each of these datasets:

    a) Training set
    b) Validation set
    c) Test set

11. Ⓢ True or False: It is best practice to train a model on all of the available data, and then evaluate it on a random subset of that data.

12. True or False: You can expect that model performance will improve over time.

13. Ⓢ True or False: Consider a supervised learning model that has been trained on a dataset. If new patterns emerge in future data, the model will not be able to use them effectively.

14. What is the purpose of a loss function in machine learning?

15. Ⓢ A data scientist has developed a model to predict personal bankruptcy. She trained the model on credit scores between 500 and 820, with a mean of 700. Over the next year, the credit scores were between 450 and 750, with a mean of 650. Should the model be retrained? Explain your answer.

16. Consider a loss function that computes the difference between the target y and the prediction $\hat{y}$ as $y - \hat{y}$. Would this be a useful loss function? Explain your reasoning.

17. Ⓢ What are the strengths and weakness of the mean squared error compared to the mean absolute error?

18. You wish to run hyperparameter tuning on a hyperparameter which is real-valued. You are considering two approaches, where the first randomly selects values for evaluation, and the second uses grid search. What are the strengths and weaknesses of each approach?

19. What is the cost and benefit of increasing the number of folds in k-fold cross validation?

20. When measuring and reporting model performance, why isn't cross-entropy loss typically used?

21. Ⓢ For the binary classification problem, is precision sufficient in measuring model performance? Explain your reasoning.

22. A certain diagnostic test produces 25 true positives for each false positive. What is the precision of this test?

23. Ⓢ True or False: A recall of 80% is considered excellent for any machine learning problem.

24. A binary classifier has correctly predicted a target of 1 for each instance where the actual target was 1.

25. For the confusion matrix below, calculate the accuracy, recall, precision, and F1 score.

predicted actual	P	N
P	102	7
N	12	65

13

Modeling with Linear Regression

We have learned how to clean and prepare data, how to summarize and visualize it, and how to transform it into useful predictors. We just learned about the building blocks of machine learning. In this chapter, we fit supervised learning models to data when the target variable is continuous. This is the regression task, and we will use *linear regression* for solving the problem.

The chapter will begin with a discussion of the important concepts and mathematics, and continue with Python code for implementation. We will use `sklearn` for the modeling. As this chapter provides an outline, further exploration is encouraged. Additional resources can be found in the course repo in this folder:

`semester1/week_13_14_model/`

Tip: Be sure to confirm the validity of any data before training a model on it. Describing the data with statistical summaries and visualizations will be helpful here.

13.1 Mathematical Framework

Returning to our labeled dataset $\{(\mathbf{x}_i, y_i)\}_{i=1}^{n}$, we want to learn a function that can predict the target y given $\mathbf{x}$. A linear regression model specifies a linear relationship between $\mathbf{x}$ and the target y. This is a simpler structure than a non-linear model, which will make it easier to explain and fit to data. The tradeoff is that for a complex relationship between predictors and target, the linear model might not fit as well. Later in this chapter, we will state precisely what we mean by the model fitting well. For some intuition, consider a single predictor X, its associated response Y, and a set of observations $(x_1, y_1), (x_2, y_2), \ldots, (x_n, y_n)$. A linear model that best fits the data will minimize its distance from the points, in effect "passing through" a majority of them as in Figure 13.1.

For each observation i, the error in measuring the target is represented as ϵ_i. A common assumption made in linear regression is that the errors follow a normal distribution with mean zero and variance $\sigma^2 > 0$. This is often written as

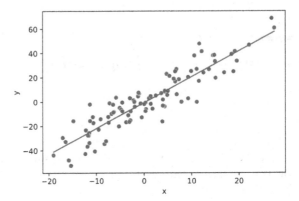

FIGURE 13.1: Scatter Plot with Line of Best Fit

$$\epsilon_i \sim \mathcal{N}(0, \sigma^2)$$

The normality assumption is not necessary for fitting the model, but it is a requirement in the statistical testing.

Next, let's look at the mathematical form for linear regression, which is stated for each observation i and p predictors:

$$y_i = \beta_0 + \beta_1 x_{i1} + \ldots + \beta_p x_{ip} + \epsilon_i, \quad i = 1, \ldots, n$$

where

- y_i is the target for observation i

- β_0 is an intercept parameter

- $\beta_1, \ldots, \beta_p$ are slope parameters or *weights*

- $x_{i1}, \ldots, x_{ip}$ are predictors for observation i

- ϵ_i is the error for observation i

In this setup, we assume that the target can be modeled by a linear combination of the predictors. There is a linear equation for each observation, for a total of n equations. There are $p+1$ unknowns, represented by the weights β_0, $\beta_1, \ldots, \beta_p$ applied to each predictor. The weighted combination of predictors is an attempt to match the target value.

As a small example, imagine we have collected three observations $(1, 2), (2, 3), (3, 5)$. There is one predictor, and the first observation has predictor value 1 and target value 2. Given this data, we can write three linear equations with two unknowns:

$$2 = \beta_0 + \beta_1(1) + \epsilon_1$$
$$3 = \beta_0 + \beta_1(2) + \epsilon_2$$
$$5 = \beta_0 + \beta_1(3) + \epsilon_3$$

Now let's guess that $\beta_0 = 0$ and $\beta_1 = 2$. This produces:

$$2 = 0 + 2 + \epsilon_1$$
$$3 = 0 + 4 + \epsilon_2$$
$$5 = 0 + 6 + \epsilon_3$$

For the first equation, the weighted combination of predictors exactly matches the target. For the next two equations, however, there are errors. In fact, this system has more equations than unknowns, and it cannot be exactly solved. It is called an *overdetermined system*. An overdetermined system is better than an *underdetermined system*, which results when there are more unknowns than equations. Underdetermined systems cannot be solved.

In algebra class, you might have seen many systems where the number of equations matched the number of unknowns. These systems can be exactly solved, which is wonderful. In practice, it won't generally be possible to have a matching number of equations and unknowns; we will collect observations which generally outnumber the predictors: $n > p$. To solve the overdetermined system, we will establish a metric and an objective, such as minimizing squared errors. This procedure will yield estimates $\hat{\beta}_i$ for each parameter β_i. Then the predictions can be estimated with a fitting equation:

$$\hat{y}_i = \hat{\beta}_0 + \hat{\beta}_1 x_{i1} + \ldots + \hat{\beta}_p x_{ip}$$

In the case of a single predictor ($p = 1$), we can drop the subscript on the predictor and write the fitting equation as:

$$\hat{y}_i = \hat{\beta}_0 + \hat{\beta}_1 x_i$$

This model is called *simple linear regression*. In the case of multiple predictors ($p > 1$), the model is sometimes called *multiple linear regression*. The prediction $\hat{y}_i$ is called the *fitted value*. The difference between the target and the fitted value, $y_i - \hat{y}_i$, is called the *residual*. The residual is an estimate of the true, unobservable error.

A common technique for fitting a linear regression model is called *ordinary least squares regression* (OLS regression), and this is what will be used here.

A squared loss function is applied to the residuals. The objective of OLS regression is to establish parameter estimates $\hat{\beta}_0, \hat{\beta}_1, \ldots, \hat{\beta}_p$ that minimize the sum of squared residuals SS_{res} defined as:

$$SS_{res}(\boldsymbol{\beta}) = \sum_{i=1}^{n} \left(y_i - (\beta_0 + \beta_1 x_{i1} + \ldots + \beta_p x_{ip})\right)^2 \qquad (13.1)$$

Minimizing $SS_{res}(\boldsymbol{\beta})$ with respect to $\boldsymbol{\beta}$ can be done with differential calculus, and this will yield a parameter estimate $\hat{\beta}_i$ for each parameter β_i. Next, we will review the details for the simple linear regression case.

13.1.1 Parameter Estimation for Linear Regression

Let us start with the residual sum of squares for the case of a single predictor:

$$SS_{res}(\boldsymbol{\beta}) = \sum_{i=1}^{n} \left(y_i - (\beta_0 + \beta_1 x_i)\right)^2$$

There are two parameters in this model: β_0 and β_1. Our strategy is to simplify the equation and look for a critical point. This is done by taking the partial derivative with respect to each parameter (i.e., computing the gradient), setting the equations to zero, and solving the system of equations.

$$SS_{res}(\boldsymbol{\beta}) = \sum_{i=1}^{n} \left(y_i^2 - 2y_i(\beta_0 + \beta_1 x_i) + (\beta_0 + \beta_1 x_i)^2\right)$$
$$= \sum_{i=1}^{n} \left(y_i^2 - 2y_i\beta_0 - 2y_i\beta_1 x_i + \beta_0^2 + 2\beta_0\beta_1 x_i + \beta_1^2 x_i^2\right)$$

Next, take the partial derivative with respect to β_0 and set equal to zero:

$$\frac{\partial SS_{res}(\boldsymbol{\beta})}{\partial \beta_0} = \sum_{i=1}^{n} \left(-2y_i + 2\beta_0 + 2\beta_1 x_i\right)$$

$$0 = -\sum_{i=1}^{n} y_i + n\hat{\beta}_0 + \hat{\beta}_1 \sum_{i=1}^{n} x_i \quad \text{(dividing through by 2)}$$

$$0 = -\bar{y} + \hat{\beta}_0 + \hat{\beta}_1 \bar{x} \qquad\qquad \text{(dividing through by } n\text{)}$$

We used the definition of the mean in the final line: $\bar{z} = \sum_{i=1}^{n} z_i / n$.

Now solve for the parameter estimate $\hat{\beta}_0$:

$$\hat{\beta}_0 = \bar{y} - \hat{\beta}_1 \bar{x}$$

Next, we need to determine the parameter estimate $\hat{\beta}_1$. We take the partial derivative of S with respect to β_1 and set equal to zero:

$$\frac{\partial SS_{res}(\beta)}{\partial \beta_1} = \sum_{i=1}^{n} \left(-2x_i y_i + 2\beta_0 x_i + 2\beta_1 x_i^2 \right)$$

$$0 = \sum_{i=1}^{n} \left(-x_i y_i + \hat{\beta}_0 x_i + \hat{\beta}_1 x_i^2 \right) \qquad \text{(dividing through by 2)}$$

$$0 = \sum_{i=1}^{n} \left(-x_i y_i + (\bar{y} - \hat{\beta}_1 \bar{x})x_i + \hat{\beta}_1 x_i^2 \right) \qquad \text{(substituting for } \hat{\beta}_0\text{)}$$

$$0 = -\sum_{i=1}^{n} x_i y_i + \sum_{i=1}^{n} x_i \bar{y} - \hat{\beta}_1 \sum_{i=1}^{n} \bar{x} x_i + \hat{\beta}_1 \sum_{i=1}^{n} x_i^2 \quad \text{(applying sums)}$$

$$0 = -\sum_{i=1}^{n} x_i(y_i - \bar{y}) + \hat{\beta}_1 \sum_{i=1}^{n} x_i(x_i - \bar{x}) \qquad \text{(grouping terms)}$$

Now solve for the parameter estimate $\hat{\beta}_1$:

$$\hat{\beta}_1 = \frac{\sum_{i=1}^{n} x_i(y_i - \bar{y})}{\sum_{i=1}^{n} x_i(x_i - \bar{x})}$$

This is equivalent to the common equation for $\hat{\beta}_1$, which is left as an exercise:

$$\hat{\beta}_1 = \frac{\sum_{i=1}^{n}(x_i - \bar{x})(y_i - \bar{y})}{\sum_{i=1}^{n}(x_i - \bar{x})^2}$$

We have now seen how to derive the simple linear regression parameter estimates using differential calculus. This is done in software, but it is a valuable exercise nonetheless. For the more general case, we can use matrices. This will handle the simple linear regression case that we just completed, as well as the multiple linear regression case. This involves creating:

- A *design matrix* $\mathbf{X}$ which holds the predictor data. The n rows represent the observations and the $p + 1$ columns represent the intercept term (in the first column) and the p predictors in the remaining columns. For example, element x_{12} is the value of predictor 2 for observation 1.

$$\mathbf{X} = \begin{bmatrix} 1 & x_{11} & x_{12} & \cdots & x_{1p} \\ 1 & x_{21} & x_{22} & \cdots & x_{2p} \\ \vdots & \vdots & \vdots & \ddots & \vdots \\ 1 & x_{n1} & x_{n2} & \cdots & x_{np} \end{bmatrix}$$

- Vectors for the parameters $\boldsymbol{\beta}$, targets $\mathbf{y}$, and errors $\boldsymbol{\epsilon}$:

$$\boldsymbol{\beta} = \begin{bmatrix} \beta_0 \\ \beta_1 \\ \vdots \\ \beta_p \end{bmatrix} \qquad \mathbf{y} = \begin{bmatrix} y_1 \\ y_2 \\ \vdots \\ y_n \end{bmatrix} \qquad \boldsymbol{\epsilon} = \begin{bmatrix} \epsilon_1 \\ \epsilon_2 \\ \vdots \\ \epsilon_n \end{bmatrix}$$

Putting the pieces together allows for writing the system of equations in compact matrix notation:

$$\mathbf{y} = \mathbf{X}\boldsymbol{\beta} + \boldsymbol{\epsilon}$$

In theory, it is possible to solve this system of equations analytically for the parameter vector. This yields the following equation, which is derived in this chapter's appendix:

$$\hat{\boldsymbol{\beta}} = (\mathbf{X}^T\mathbf{X})^{-1}\mathbf{X}^T\mathbf{y}$$

The notation $\mathbf{X}^T$ and $(\mathbf{X}^T\mathbf{X})^{-1}$ denotes the transpose of $\mathbf{X}$ and inverse of $\mathbf{X}^T\mathbf{X}$, respectively. In practice, computation of the inverse may be intractable for massive datasets. In this case, an iterative approach such as gradient descent (GD) can be followed. We will discuss GD in the next chapter on logistic regression. Software packages for big data such as Apache Spark will follow the iterative approach.

Given the parameter estimates, the fitted values can be computed. The fitting question can be written in matrix form as follows:

$$\hat{\mathbf{y}} = \mathbf{X}\hat{\boldsymbol{\beta}}$$

The matrix notation aligns with how we organize data for computation. A `pandas` dataframe will hold the data in a NumPy array which arranges the observations on the rows and the predictors on the columns. When fitting a model in `sklearn`, the matrix of predictors $\mathbf{X}$ and the vector of targets $\mathbf{y}$ will be passed to a function. It is not necessary to include the column of `1s` in $\mathbf{X}$.

Now that we have briefly covered the conceptual and mathematical framework of the linear regression model, we can think about predictors and code implementation. For a deeper dive into linear regression, [39] and [40] are excellent references. In the next section, we will briefly learn some qualitative best practices for predictor selection.

13.2   Being Thoughtful about Predictors

When selecting model predictors, it may be tempting to include all available variables in the model. It is certainly possible to use an algorithm to automate this step. However, a word of caution is in order, as there are some qualitative considerations when deciding predictors:

- They should make intuitive sense. In practice, the data scientist might discuss the predictors with the line of business, the product manager, and other stakeholders.

- They should not introduce bias or ethical issues into the model

One needs to be careful around the use of demographic variables, as it can be easy to discriminate based on age, gender, race, and ethnicity. Examples of variables that would introduce bias, and therefore should be avoided in modeling, include:

- Age, race, ethnicity, gender, religion, disability, color, national origin, religion

- Employment type, where jobs like nursing can introduce a gender bias

- An indicator if a caller pressed 2 to hear a message in Spanish, which can introduce an ethnicity bias

- Salary, which can introduce a race/ethnicity bias

In summary, the model predictors should be carefully considered. It is not enough that predictors improve the model fit; they must also make sense and not cause harm.

13.3   Predicting Housing Prices

In this exercise, we will fit a linear regression model to a dataset of California housing prices from the StatLib repository at Carnegie Mellon. It was originally featured in [41]. The data is from the 1990 U.S. census and includes variables such as population, median income, and median house price for each block group. A block group is the smallest geographical unit for published census sample data (typically 600 to 3000 people). We will begin by importing relevant modules and exploring the data in Figure 13.2.

```
from sklearn.datasets import fetch_california_housing
from sklearn.model_selection import train_test_split
from sklearn.preprocessing import StandardScaler
from sklearn.linear_model import LinearRegression
from sklearn.metrics import r2_score # R-squared

import pandas as pd

housing = fetch_california_housing()
```

```
[5]: {'data': array([[  8.3252   ,  41.      ,  6.98412698, ...,  2.55555556,
          37.88     , -122.23   ],
       [  8.3014   ,  21.      ,  6.23813708, ...,  2.10984183,
          37.86     , -122.22   ],
       [  7.2574   ,  52.      ,  8.28813559, ...,  2.80225989,
          37.85     , -122.24   ],
       ...,
       [  1.7      ,  17.      ,  5.20554273, ...,  2.3256351 ,
          39.43     , -121.22   ],
       [  1.8672   ,  18.      ,  5.32951289, ...,  2.12320917,
          39.43     , -121.32   ],
       [  2.3886   ,  16.      ,  5.25471698, ...,  2.61698113,
          39.37     , -121.24   ]]),
 'target': array([4.526, 3.585, 3.521, ..., 0.923, 0.847, 0.894]),
 'frame': None,
 'target_names': ['MedHouseVal'],
 'feature_names': ['MedInc',
  'HouseAge',
  'AveRooms',
  'AveBedrms',
  'Population',
  'AveOccup',
  'Latitude',
  'Longitude'],
 'DESCR': '.. _california_housing_dataset:\n\nCalifornia Housing dataset\n-------------
-------------\n\n**Data Set Characteristics:**\n\n    :Number of Instances: 20640\n\n
 :Number of Attributes: 8 numeric, predictive attributes and the target\n\n    :Attribut
e Information:\n        - MedInc        median income in block group\n        - HouseAg
e      median house age in block group\n        - AveRooms      average number of rooms
per household\n        - AveBedrms     average number of bedrooms per household\n
- Population    block group population\n        - AveOccup      average number of house
```

FIGURE 13.2: California Housing Data Snapshot

Checking the data type with `type(housing)` shows this specialized object: `sklearn.utils.Bunch`. As the object appears to contain key:value pairs, we request the keys:

```
housing.keys()
```

OUT:

```
dict_keys(['data', 'target', 'frame', 'target_names',
 'feature_names', 'DESCR'])
```

The key DESCR provides descriptive data about the dataset. The name of the target variable is MedHouseVal, which is the median house value in units of $100K. The first two target values are 4.526 and 3.585. Predictors include the age of the house (`HouseAge`), average number of rooms (`AveRooms`), and average number of bedrooms (`AveBedrms`). Both the target values and the predictor values are stored in NumPy arrays.

Let's check the shape of the predictor data:

```
housing['data'].shape
```

```
OUT:
returns (20640, 8)
```

There are 20640 rows, or block groups, and 8 columns containing predictors. Note that for this data type, we can reference an attribute like `data` in two ways:

```
housing['data']
```

or

```
housing.data
```

13.3.1   Data Splitting

In practice, we would spend some additional time exploring the data to understand its properties and weaknesses. For brevity, let's assume this was done. Next, since we want to train a model and assess how its performance generalizes to unseen data, we will split the dataset into a training set and a test set. We will train the model on the training set, and we can evaluate its performance on both the training set and the test set. We expect performance to be better on the former, as the model will learn patterns in this set. The hope is that the model performs similarly on the test set. For splitting the data, we will use the function `train_test_split()` from `sklearn`. It takes four parameters and returns four datasets. Let's look at the code and then a description of how it works:

```
x_train, x_test, y_train, y_test =
  train_test_split(housing.data,
                   housing.target,
                   train_size = 0.6,
                   random_state=314)
```

INPUTS:

- `housing.data` – the predictor data
- `housing.target` – the target data

- `train_size` – the fraction of data used in the training set (the remainder is used in the test set)

- `random_state` – a seed for replicating the random split of rows into train and test sets

OUTPUTS:

- `x_train` – the predictor training data

- `x_test` – the predictor test data

- `y_train` – the target training data

- `y_test` – the target test data

The `train_test_split()` function split the data into 60% train and 40% test, producing four new NumPy arrays. We can check the number of records in each piece by computing their lengths:

```
len(x_train), len(x_test), len(y_train), len(y_test)
OUT:
```

```
(12384, 8256, 12384, 8256)
```

Given 12384 rows in the training set, this represents $12384/(12384+8256) = 60\%$ of the total rows, which is correct.

13.3.2   Data Scaling

Next, we will scale the training data and the test data using `StandardScaler()`. This is done after splitting because the predictor standard deviations need to be measured on the training set. The test set is treated as an independent data set, and it would not be correct to use that data in the standard deviation measurements.

The code snippet below will instantiate a `StandardScaler` object and scale the training data. Specifically, the mean and standard deviation will be computed for each column. Next, each column will have its mean subtracted, and the result will be divided by the column standard deviation.

```
scaler = StandardScaler()
x_train_s = scaler.fit_transform(x_train)
```

We can inspect the scaled data:

```
x_train_s
```

```
array([[-5.41768936e-01,  8.97274139e-01, -1.85640214e-02, ...,
        -4.28501066e-02, -2.20840635e-01,  5.77185876e-02],
       ...,
        -6.70406776e-02, -1.40562124e+00,  1.25411953e+00]])
```

the column means:

```
x_train_s.mean(axis=0)
```

OUT:

```
array([-4.39298549e-15, -1.02774522e-16, -1.06658763e-14,
       -5.55920742e-15, -1.52942545e-17, -7.05635754e-16,
        9.44804814e-15, -8.41492753e-14])
```

and the column standard deviations:

```
x_train_s.std(axis=0)
```

OUT:
```
array([1., 1., 1., 1., 1., 1., 1., 1.])
```

The column means are approximately zero and the standard deviations are one, as we expect.

13.3.3   Model Fitting

Next, we train (or fit) a linear regression model on the training data. OLS regression can be implemented with the `LinearRegression()` function from `sklearn`. We can instantiate the model and fit it to training data in a single line. By default, an intercept term is included. The order of the parameters is important: the `fit()` function takes X followed by y as in this code snippet:

```
reg = LinearRegression().fit(x_train_s, y_train)
```

The `reg` object holds the trained regression model, and we can explore its contents. We can list its attributes and methods like this:

```
dir(reg)
```

OUT:
```
['__abstractmethods__',
 '__class__',
 '__delattr__',
 ...,
```

```
'fit',
'fit_intercept',
'get_params',
'intercept_',
...,
'score',
'set_params',
'singular_']
```

Among the list are methods for fitting the model with `fit()` and extracting the intercept estimate. Next, let's extract the intercept estimate:

```
reg.intercept_
```

```
OUT:
2.060088804909497
```

and slope estimates:

```
reg.coef_
```

```
OUT:
array([ 0.81403101,  0.11883616, -0.260123  ,  0.31025271,
       -0.00178077, -0.04600269, -0.91689468, -0.88930004])
```

Let's code the mathematical formula for the parameter estimates. Then we can verify that the calculated results match the estimates from **sklearn**. As a reminder, here is the formula:

$$\hat{\beta} = (\mathbf{X}^T\mathbf{X})^{-1}\mathbf{X}^T\boldsymbol{y}$$

In the code snippet, the @ symbol performs matrix multiplication:

```
import numpy as np

# in design matrix, prepend column of ones for the
# intercept term
# hstack() horizontally stacks the two arrays
X = np.hstack((np.ones((x_train_s.shape[0], 1)), x_train_s))

XtX = X.T @ X

# compute matrix inverse
XtXi = np.linalg.inv(XtX)

betas = XtXi @ (X.T @ y_train)
betas
```

```
OUT
array([ 2.06008880e+00,  8.14031008e-01,  1.18836164e-01,
       -2.60123003e-01, 3.10252714e-01, -1.78076796e-03,
       -4.60026851e-02, -9.16894684e-01, -8.89300037e-01])
```

The vector includes the intercept term as the first element, and the slopes as the remaining elements. This matches the estimates from `sklearn`. Generally, we won't need to explicitly use the formula, but it is good to understand what the software is doing. Next, let's bring up a list of the predictors to understand what the slopes mean:

```
housing.feature_names
```

```
OUT:
['MedInc',
 'HouseAge',
 'AveRooms',
 'AveBedrms',
 'Population',
 'AveOccup',
 'Latitude',
 'Longitude']
```

As the order of the predictors corresponds with the slope estimates, the 'MedInc' variable (median income in the block group) has value 0.814 after rounding. The positive sign indicates that higher median income is associated with a higher median house value, on average. The parameter estimates can be interpreted as the incremental change in target value for an incremental change in the predictor. For each additional unit in median income in the block group, the increase in median house value would be 0.814 x 100K = $81,400, on average, assuming all other predictors are held constant.

The regression intercept estimate of 2.06, after rounding, indicates the average target value when all of the predictor values are zero. This is only meaningful if it's reasonable for all predictor values to actually be equal to zero. In this case, it does not make sense for the average number of rooms and bedrooms to be zero, and so the intercept interpretation is not useful.

Next, let's use the model to make predictions on the test set. First, we will need apply the `StandardScaler` to scale the data:

```
x_test_s = scaler.fit_transform(x_test)
```

Next, we run the scaled test data through the model to predict target values $\hat{y}$. The `predict()` function is used for the fitting equation $\hat{y} = \mathbf{X}\hat{\beta}$.

```
y_test_predicted = reg.predict(x_test_s)
```

13.3.4   Interpreting the Parameter Estimates

The `sklearn` package allowed for fitting the model and working with the parameter estimates, but it didn't tell us which estimates, if any, are statistically different from zero. The `statsmodels` package can fit the same linear regression model and run one-sample t-tests to answer this question. Since we have already done the work of preparing the data, the code is minimal:

```
# import statsmodels
import statsmodels.api as sm

# append column of 1s to design matrix for intercept term
X = sm.add_constant(x_train_s)

# fit the OLS model
results = sm.OLS(y_train, X).fit()

# print the model summary
print(results.summary())
```

Table 13.1 shows the parameter estimate table from the model summary. The p-values and confidence intervals indicate that all variables except $x5$ ('Population') are significant. For $x5$, the p-value is much greater than 0.05, and zero is contained in the 95% confidence interval. The next step would be to drop this variable and refit the model.

TABLE 13.1: Housing Parameter Estimate Table

	coef	std err	t	$P > \lvert t \rvert$	[0.025	0.975]
const	2.0601	0.007	315.640	0.000	2.047	2.073
x1	0.814	0.010	79.647	0.000	0.794	0.834
x2	0.1188	0.007	16.417	0.000	0.105	0.133
x3	−0.2601	0.020	−12.777	0.000	−0.300	−0.220
x4	0.3103	0.019	16.379	0.000	0.273	0.347
x5	−0.0018	0.007	−0.256	0.7998	−0.015	0.012
x6	−0.0460	0.007	−7.006	0.000	−0.059	−0.033
x7	−0.9169	0.020	−46.580	0.000	−0.955	−0.878
x8	−0.8893	0.019	−46.003	0.000	−0.927	−0.851

13.3.5   Model Performance Evaluation

Since we know the target values from the test set, `y_test`, we can compare them to our predictions to assess model performance. We will use R^2 as the metric. As a reminder, R^2 measures the fraction of variation in Y explained by the predictors. The details of the R^2 calculation are included at the end of this section.

The r2_score() function from sklearn will compute R^2. It takes an array of actual target values and an array of predicted target values, in that order.

```
print(r2_score(y_test, y_test_predicted))
```

```
OUT:
0.6078399673783987
```

The R^2 indicates that 61% of the variation in the median house value is explained by the predictors. The set of predictors is useful, but there is a large fraction of unexplained variation. It can help to plot the data to understand large discrepancies between actual and predicted values, so let's make a scatterplot of these quantities.

```
import seaborn as sns
import matplotlib.pyplot as plt

sns.scatterplot(x=y_test, y=y_test_predicted)
plt.plot((-2,7),(-2,7))
plt.xlabel('actual')
plt.ylabel('predicted')
```

Overplotted on the predicted versus actual graph in Figure 13.3 is a 45-degree line.[1] Ideally, the data points would fall on this 45-degree line, indicating a strong correlation between actual and predicted values. This pattern is roughly followed, but there are some odd effects to discuss. The first is the vertical stack of points at the far right of the plot. This reflects the actual target reaching a maximum of roughly five, while the predicted values range from roughly one to seven. The model is missing the mark on many of these higher-valued block groups. Second, some of the predictions are negatively valued, which does not make sense. The lowest point on the plot, where the actual value is near four while the prediction is near negative two, is one obvious example. This point might represent bad data or an unusual pattern, and it can be explored. It can also be helpful to review the observations with the largest absolute difference between the prediction and target.

A common diagnostic plot shows the residuals on the y-axis and the predicted (fitted) values on the x-axis as in Figure 13.4. A horizontal line at zero helps understand if the residuals cluster about this line for different predictor values. This is important, as one of the assumptions is that the error term

---

[1]This is done by finding the minimum and maximum values over all actual and predicted values, and passing (min_val, max_val), (min_val, max_val) to the plot() function. The ordered pairs represent the x-coordinates and y-coordinates, respectively.

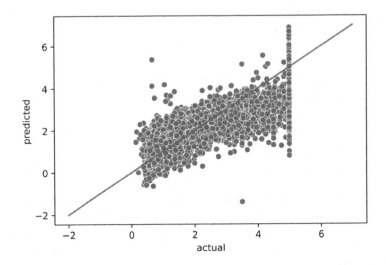

FIGURE 13.3: California Housing: Predicted versus Actual Targets

has mean zero. There should not be patterns in the data. The presence of a trend line or curve would be problematic, as this would suggest one or more missing predictors. The residual plot here is problematic, as the residuals do not cluster about the zero line, and there is a negatively sloping set of points. The sloping points are a result of the target maximum value of five.

At this stage, an analyst may spend a fair bit of time doing the mentioned error analysis. Creating various plots and filtering and sorting the data in different ways will provide insight. Defects in the data may be discovered and fixed, such as incorrect predictor or target values. When updates need to be made to the data, it is best to make the changes in code and save the adjusted data to separate files. This will make the changes transparent and repeatable, and it will be possible to undo the changes if necessary. For further model improvement, there may be additional predictors that can be engineered and included. It may also be possible to collect more data, and focus should be given where the model is weak.

13.3.6   Comparing Models

It will be important to try different models and compare their performance. Graphical methods and metrics are valuable for making these comparisons. Since we have used all of the provided predictors in building the model, let's build a second model that includes a subset of the predictors. We will select the predictor subset and then repeat the earlier steps. This code snippet selects the first three predictors from the training data:

```
x_train_subset = x_train[:,:3]
```

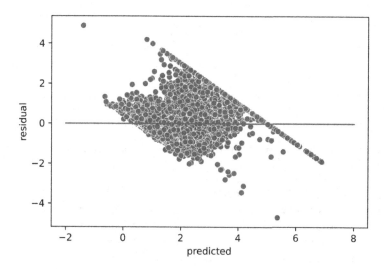

FIGURE 13.4: California Housing: Residuals versus Predicted

The notation [:,:3] references the rows and columns, respectively. The colon denotes selection of all rows, while :3 denotes selection of columns zero through two, inclusive.

Since we are changing the training data, we need to create a new StandardScaler object and transform the data:

```
scaler2 = StandardScaler()
x_train_subset_s = scaler2.fit_transform(x_train_subset)
```

Next, we create a new LinearRegression object and fit it to the data:

```
reg2 = LinearRegression().fit(x_train_subset_s, y_train)
```

We now have our new model. To assess the performance, we will first subset the test set and scale it.

```
x_test_subset_s = scaler2.fit_transform(x_test[:,:3])
```

Next, we use the reg2 model to make predictions on the test set:

```
y_test_predicted2 = reg2.predict(x_test_subset_s)
```

The R^2 is useful for measuring the fraction of variance explained by the predictors, but when comparing models, the *adjusted* R^2 is more appropriate. The adjusted R^2 is a modification of R^2 that penalizes model complexity, which increases as additional parameters are included in a model. It is calculated as follows:

$$adjusted \ R^2 = 1 - (1 - R^2) \left(\frac{n-1}{n-p-1} \right)$$

where

- n is the number of observations

- p is the number of model parameters, excluding the intercept

`sklearn` doesn't have a function to compute adjusted R^2, so we can define it ourselves:

```
def adj_rsquared(X, y, y_pred):
    r2 = r2_score(y, y_pred)   # compute R-squared
    n = len(y)                 # number of observations
    p = X.shape[1]             # number of parameters
    adj_r2 = 1 - (1 - r2) * (n - 1)/(n - p - 1)
    return adj_r2
```

Next, let's compute the R^2 and the adjusted R^2 for this smaller model:

```
print('R2:', r2_score(y_test, y_test_predicted2))

print('adj R2:', adj_rsquared(x_test_subset_s,
                              y_test,
                              y_test_predicted2))
```

```
OUT:
R2:      0.5206094971517826
adj R2:  0.5204352155826424
```

For comparison, we compute the R^2 and adjusted R^2 for the larger model:

```
print('R2:', r2_score(y_test, y_test_predicted))

print('adj R2:', adj_rsquared(x_test_s,
                             y_test,
                             y_test_predicted))
```

```
OUT:
R2:      0.6078399673783987
adj R2:  0.6074595526505009
```

There are two important observations. First, the smaller model has a lower adjusted R^2 than the larger model. This tells us that the predictors we dropped are important in explaining the target. Second, for each model, the adjusted

R^2 is very close to the R^2. If we look at the formula for adjusted R^2, we notice an adjustment factor of $(n-1)/(n-p-1)$. As n increases relative to p, the factor approaches one, which implies the adjusted R^2 approaches R^2. For this example, $n = 8256$, $p_{small} = 3$, and $p_{large} = 8$. The adjustment factor for the small model and large model is $(8256 - 1)/(8256 - 3 - 1) = 1.00036$ and $(8256 - 1)/(8256 - 8 - 1) = 1.00097$, respectively. There is a greater adjustment for the larger model due to its higher complexity, but in each case, the adjustments are very small.

13.3.7 Calculating R-Squared

R^2 is easily calculated in software, but it is important to understand its conceptual meaning and calculation. Conceptually, we state an equality about the total sum of squares which is proportional to the variance in the target data:

total sum of squares = explained sum of squares + residual sum of squares

denoting

- SS_{tot} = total sum of squares

- SS_{exp} = explained sum of squares

- SS_{res} = residual sum of squares

- $R^2 = SS_{exp}/SS_{tot}$

gives

$$SS_{tot} = SS_{exp} + SS_{res}$$

Dividing through by SS_{tot} gives:

$$1 = \frac{SS_{exp}}{SS_{tot}} + \frac{SS_{res}}{SS_{tot}}$$

Substituting for R^2 gives:

$$1 = R^2 + \frac{SS_{res}}{SS_{tot}}$$

Lastly, we solve for R^2:

$$R^2 = 1 - \frac{SS_{res}}{SS_{tot}}$$

This gives a compact formula for R^2 and a nice interpretation: a model that explains a large fraction of the variation in the target variable will produce a small residual sum of squares relative to the total sum of squares, SS_{res}/SS_{tot}.

In turn, this will result in a large R^2. If the model perfectly fits the data, SS_{res} will be zero and R^2 will be one. If the model does not explain any of the variation in the target, SS_{res} will equal SS_{tot} and $R^2 = 1 - 1 = 0$. Next, we need to explain exactly how to compute SS_{res} and SS_{tot}.

As shown earlier, SS_{res} is computed by:

1. Calculating residuals $y_i - \hat{y}$ for each data point i
2. Squaring the residuals
3. Summing the squared residuals

As a formula,

$$SS_{res} = \sum_{i=1}^{n} (y_i - \hat{y})^2$$

SS_{tot} is computed by:

1. Computing the average of the data, $\bar{y}$
2. Calculating deviations $y_i - \bar{y}$ for each data point i
3. Squaring the deviations
4. Summing the squared deviations

As a formula,

$$SS_{tot} = \sum_{i=1}^{n} (y_i - \bar{y})^2$$

Again, it is generally not necessary to do these calculations by hand or spreadsheet, as R^2 is implemented in software. However, the conceptual understanding and reasoning with sums of squares is valuable.

13.4   Chapter Summary

Linear regression can be considered for supervised learning when the target variable is continuous. It uses the strong assumption of a linear relationship between predictors and the target, but this is frequently useful in practice. Simple linear regression uses a single predictor, while multiple linear regression uses more than one predictor to explain the target.

To put the model to work, the parameters need to be estimated from the data. We used ordinary least squares regression for estimation, which selects parameter values minimizing the sum of squared differences between targets and predictions. The differences are called residuals.

Variables used as predictors should be intuitive and they should avoid introducing bias and ethical issues into the model. While demographic variables might increase explanatory power, they can systematically discriminate against subpopulations.

We fit a linear regression model to California housing prices to predict median home price. During the exercise, we split the data, scaled the predictors, fit the model, and assessed performance. We measured the R^2 on the test set, which represents the fraction of variance in the target explained by the model. Additionally, we produced a residual plot and a fitted-versus-actual plot to diagnose the model. The residual plot should show no patterns and a random scatter about zero, but in fact, this data showed a pattern and was far from random. This suggests more work ahead, such as including additional predictors. Lastly, we fit a second model to the data and compared model performance with adjusted R^2. This metric penalizes for additional complexity, making it appropriate for comparing regression models of different size.

In the next chapter, we will work with data to predict a target which can take two possible values. This is the binary classification problem, and we will learn a very useful and prevalent model for tackling it: *logistic regression*.

13.5 Appendix: Parameter Estimation in Matrix Form

We return to the system of equations in compact matrix notation:

$$y = \mathbf{X}\beta + \epsilon$$

Next, we write the equation for the sum of squared residuals:

$$SS_{res}(\beta) = (y - \mathbf{X}\beta)^T(y - \mathbf{X}\beta)$$

Now we take the derivative of SS_{res} with respect to β. This is like taking the derivative of quadratic function $(y - \mathbf{X}\beta)^2$. The result is 2 times the derivative of $(y - \mathbf{X}\beta)^T$ with respect to β:

$$\frac{dSS_{res}(\beta)}{d\beta} = -2\mathbf{X}^T(y - \mathbf{X}\beta)$$

$$0 = -2\mathbf{X}^T(y - \mathbf{X}\beta) \quad \text{(set equal to zero)}$$

Note that $\mathbf{0}$ is the zero vector. Next, we rearrange:

$$\mathbf{X}^T \boldsymbol{y} = \mathbf{X}^T \mathbf{X} \boldsymbol{\beta}$$

This produces the *normal equations*. The last step is to solve for $\boldsymbol{\beta}$. We multiply on the left by the inverse of the matrix $\mathbf{X}^T \mathbf{X}$. This is done to each side of the equation:

$$\hat{\boldsymbol{\beta}} = (\mathbf{X}^T \mathbf{X})^{-1} \mathbf{X}^T \boldsymbol{y}$$

This section required knowledge of linear algebra and matrix derivatives. A helpful reference is [42]. The important part is to remember the matrix formula for the parameter estimates. This formula makes it possible to solve the linear regression problem for any number of parameters.

13.6   Exercises

Exercises with solutions are marked with Ⓢ. Solutions can be found in the `book_materials` folder of the course repo at this location:

https://github.com/PredictioNN/intro_data_science_course/

1. What assumptions are made in the linear regression model?

2. Ⓢ True or False: You have a system of linear equations with more equations than unknowns. This system can be solved exactly.

3. True or False: The linear regression model uses a weighted combination of predictors to closely match the target values.

4. Ⓢ True or False: An intercept term adds flexibility to the linear regression model.

5. Explain the difference between residuals and errors in the context of regression.

6. Ⓢ Explain how the parameters of a linear regression model can be estimated. The discussion should include the concept of critical points.

7. True or False: When estimating the parameters of a simple linear regression model, it is necessary to compute partial derivatives.

8. Ⓢ Consider a simple linear regression model where the predictor is uncorrelated to the response variable. What do you expect will be the value of the slope parameter?

9. How can the intercept coefficient be interpreted in a linear regression model?

10. Ⓢ True or False: You have a dataset with tens of thousands of potential predictors. The best practice is for the model to include all of the statistically significant predictors.

11. Verify the equivalence of the two forms from the simple linear regression parameter estimate solution:

$$\hat{\beta}_1 = \frac{\sum_{i=1}^n x_i(y_i - \bar{y})}{\sum_{i=1}^n x_i(x_i - \bar{x})}$$

$$\hat{\beta}_1 = \frac{\sum_{i=1}^n (x_i - \bar{x})(y_i - \bar{y})}{\sum_{i=1}^n (x_i - \bar{x})^2}$$

Hint: After expansion, you might find terms involving $\sum_{i=1}^n (x_i - \bar{x})$. Intuitively, what is the value of this expression?

12. What is a benefit of setting the seed when calling the `train_test_split()` model?

13. Ⓢ Explain how R^2 is related to correlation.

14. True or False: If the regression predictors fail to explain all of the variation in the target variable, then the model should never be used.

15. What is the purpose of a residual plot?

16. Ⓢ A residual plot shows a clear pattern. What does this mean?

17. What is a useful performance metric for comparing regression models? Explain your answer.

18. Ⓢ Using the `sklearn` module, fit a linear regression model to this dataset: $\{(3, 6), (3.5, 9), (4, 8)\}$. Show your code and print the intercept and slope coefficient estimates and the R-squared.

19. Using a spreadsheet or calculator, compute the intercept and slope coefficient estimates from a linear regression model given the dataset: $\{(3, 6), (3.5, 9), (4, 8)\}$. Organize and show all of your calculations.

14

Classification with Logistic Regression

In the last chapter, we learned a modeling approach when the target variable is continuous. In this chapter, we will learn a model for the case when the target is binary. This task is widespread across domains, such as predicting survival, the movement of a stock price, or the outcome of a sporting event. Logistic regression is an appealing model because it can work well in practice and the results are interpretable. The model is a good starting point before attempting more complex models.

We will begin by outlining the mathematical framework, which makes use of the calculus concepts we covered earlier. Following along is encouraged, but a high-level understanding will be enough. Estimation of the parameters in logistic regression cannot be done in closed form, and we will review gradient descent as an iterative solution. Next, we will build a model to detect malignancy, and learn how to interpret the parameter estimates. Python packages will do the model fitting for us, but we will confirm the results by running gradient descent. Lastly, we will evaluate model performance with binary classifier metrics. Additional resources can be found in the course repo in this folder:

`semester2/week_03_04_classification_and_logistic_regression/`

14.1   Mathematical Framework

As in the regression task, a labeled dataset $\{(\mathbf{x}_i, y_i)\}_{i=1}^n$ is available. The goal is to learn a function that can predict the target y given $\mathbf{x}$. We regard the outcome as an event that occurs ($y = 1$) with probability p and does not occur ($y = 0$) with probability $1 - p$. Given this setup, we need a function that takes predictors as inputs and produces the probability of the event as output. The predictors can potentially have domain $(-\infty, \infty)$.

A first thought might be to reuse our linear regression approach, which consists of a linear combination of the predictors. The trouble with this approach is that the range can be $(-\infty, \infty)$, which is inconsistent with probabilities. Instead, we begin with the *sigmoid function* to model probabilities. It has form:

$$p(t) = \sigma(t) = \frac{1}{1+e^{-t}}$$

An example sigmoid function is shown in Figure 14.1. The function is useful because it has domain $(-\infty, \infty)$ and range $(0, 1)$. Given this range, the output can be treated as the probability of the event. Notice the S shape and horizontal asymptotes at zero and one when the input tends to $-\infty$ and ∞, respectively.

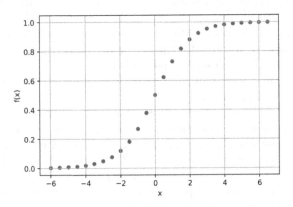

FIGURE 14.1: Sigmoid Function

The *odds* of an event is the ratio of the probability of the event taking place divided by the probability it does not take place, denoted $p/(1-p)$. We first simplify $1-p$, showing the dependence on t:

$$
\begin{aligned}
1 - p(t) &= 1 - \frac{1}{1+e^{-t}} \\
&= \frac{1+e^{-t}}{1+e^{-t}} - \frac{1}{1+e^{-t}} \\
&= \frac{e^{-t}}{1+e^{-t}}
\end{aligned}
$$

Then the odds has form

$$
\begin{aligned}
\frac{p(t)}{1-p(t)} &= \frac{1/(1+e^{-t})}{e^{-t}/(1+e^{-t})} \\
&= \frac{1}{e^{-t}} \\
&= e^{t}
\end{aligned}
$$

This expresses the odds as an exponential function of t. We can take the logarithm of both sides to yield an expression for the *log odds*:

$$ln \frac{p(t)}{1 - p(t)} = t$$

This results from $exp()$ and $ln()$ being inverses of one another. We now assume that t is a linear combination of the predictors:

$$t = \beta_0 + \beta_1 x_{i1} + \ldots + \beta_k x_{ik}$$

In fact, the right-hand side is the specification from the linear regression model. Then the equation for the log odds is

$$ln \frac{p(\mathbf{x})}{1 - p(\mathbf{x})} = \beta_0 + \beta_1 x_{i1} + \ldots + \beta_k x_{ik}$$

This is the equation for logistic regression. It assumes a linear relationship between the predictors and the log odds of the probability of the event. In the case of a single predictor ($k = 1$), we can drop the subscript on the predictor and write the equation as

$$ln \frac{p(x)}{1 - p(x)} = \beta_0 + \beta_1 x_i$$

14.1.1   Parameter Estimation for Logistic Regression

Now that the form is specified, we will need parameter estimates $\hat{\beta}_i$ for each parameter β_i where $i = 0, 1, \ldots, k$. Our technique will be to write the probability of observing a sequence of outcomes as a function of the parameters (this is called the *likelihood function*), and to apply gradient descent to maximize a version of the likelihood.

> Note: This section is more mathematically heavy than the earlier sections. It will be good to attempt the section and see the application of preliminary material, such as calculus, but it is not required.

Each outcome is a Bernoulli random variable with success probability $p(t)$, which is a function of the data and parameters. Given this information, a sequence of outcomes can be treated as a sequence of independent Bernoulli trials. The likelihood of observing the outcomes, denoted $\mathcal{L}(\boldsymbol{\beta})$, has this form:

$$\mathcal{L}(\boldsymbol{\beta}) = \prod_{i=1}^{n} p(\mathbf{x}_i|\boldsymbol{\beta})^{y_i} (1 - p(\mathbf{x}_i|\boldsymbol{\beta}))^{1-y_i}$$

The term $p(\boldsymbol{x}_i|\boldsymbol{\beta})$ denotes the probability of the event for outcome i given the parameter values, and the term $1 - p(\boldsymbol{x}_i|\boldsymbol{\beta})$ denotes the probability of a non-event for outcome i. The exponents y_i and $1 - y_i$ allow for a compact expression for a Bernoulli trial. For example, if the event occurs for outcome i, then $y_i = 1$ and $1 - y_i = 0$. It follows that

$$p(\boldsymbol{x}_i|\boldsymbol{\beta})^{y_i}(1 - p(\boldsymbol{x}_i|\boldsymbol{\beta}))^{1-y_i} = p(\boldsymbol{x}_i|\boldsymbol{\beta})^1(1 - p(\boldsymbol{x}_i|\boldsymbol{\beta}))^0$$
$$= p(\boldsymbol{x}_i|\boldsymbol{\beta})$$

which represents the probability of an event for outcome i. Similarly, if the event did not occur for outcome i, then $y_i = 0$ and $1 - y_i = 1$. It follows that

$$p(\boldsymbol{x}_i|\boldsymbol{\beta})^{y_i}(1 - p(\boldsymbol{x}_i|\boldsymbol{\beta}))^{1-y_i} = p(\boldsymbol{x}_i|\boldsymbol{\beta})^0(1 - p(\boldsymbol{x}_i|\boldsymbol{\beta}))^1$$
$$= 1 - p(\boldsymbol{x}_i|\boldsymbol{\beta})$$

This term represents the probability of a non-event for outcome i.

Lastly, the product notation $\prod_{i=1}^{n}$ expresses the sequence of Bernoulli trials. For the case of three trials ($n = 3$) all resulting in events, the likelihood will be

$$\mathcal{L}(\boldsymbol{\beta}) = \prod_{i=1}^{n} p(\boldsymbol{x}_i|\boldsymbol{\beta})^{y_i}(1 - p(\boldsymbol{x}_i|\boldsymbol{\beta}))^{1-y_i}$$
$$= \prod_{i=1}^{3} p(\boldsymbol{x}_i|\boldsymbol{\beta})$$
$$= p(\boldsymbol{x}_1|\boldsymbol{\beta})p(\boldsymbol{x}_2|\boldsymbol{\beta})p(\boldsymbol{x}_3|\boldsymbol{\beta})$$

It is easier to work with the *log-likelihood*, or the logarithm of the likelihood, as the products become sums. It is valid to apply the logarithm, as the parameters that maximize the likelihood will also maximize the log-likelihood. This is because $log()$ is a monotonically increasing function. The log-likelihood is denoted $l(\boldsymbol{\beta})$ and has the form

$$l(\boldsymbol{\beta}) = \sum_{i=1}^{n} y_i \, ln \, p(\boldsymbol{x}_i|\boldsymbol{\beta}) + (1 - y_i) \, ln \, (1 - p(\boldsymbol{x}_i|\boldsymbol{\beta}))$$

This uses two properties of logarithms:

$$log(ab) = log(a) + log(b)$$
$$log(a)^b = b \, log(a)$$

Filling in the steps,

$$l(\boldsymbol{\beta}) = ln \prod_{i=1}^{n} p(\boldsymbol{x}_i|\boldsymbol{\beta})^{y_i} (1 - p(\boldsymbol{x}_i|\boldsymbol{\beta}))^{1-y_i}$$

$$= \sum_{i=1}^{n} ln \, p(\boldsymbol{x}_i|\boldsymbol{\beta})^{y_i} (1 - p(\boldsymbol{x}_i|\boldsymbol{\beta}))^{1-y_i}$$

$$= \sum_{i=1}^{n} ln \, p(\boldsymbol{x}_i|\boldsymbol{\beta})^{y_i} + ln \, (1 - p(\boldsymbol{x}_i|\boldsymbol{\beta}))^{1-y_i}$$

$$= \sum_{i=1}^{n} y_i \, ln \, p(\boldsymbol{x}_i|\boldsymbol{\beta}) + (1 - y_i) \, ln \, (1 - p(\boldsymbol{x}_i|\boldsymbol{\beta}))$$

> Note: In the introductory chapter on machine learning, we defined the cross-entropy loss for binary classification as follows:
>
> $$H(\hat{p}_i, y_i) = -(y_i \, ln \, (\hat{p}_i) + (1 - y_i) \, ln \, (1 - \hat{p}_i))$$
>
> If we sum the cross-entropy loss over n observations, it will be equivalent to the negative log-likelihood.

We can apply the logarithm property $log(a/b) = log(a) - log(b)$ to yield

$$l(\boldsymbol{\beta}) = \sum_{i=1}^{n} ln \, (1 - p(\boldsymbol{x}_i|\boldsymbol{\beta})) + y_i \, ln \, \frac{p(\boldsymbol{x}_i|\boldsymbol{\beta})}{1 - p(\boldsymbol{x}_i|\boldsymbol{\beta})}$$

The term on the right is the log odds, which can be replaced by the linear term $\beta_0 + \beta_1 x_{i1} + \ldots + \beta_k x_{ik}$ like this:

$$l(\boldsymbol{\beta}) = \sum_{i=1}^{n} ln \, (1 - p(\boldsymbol{x}_i|\boldsymbol{\beta})) + \sum_{i=1}^{n} y_i \, (\beta_0 + \ldots + \beta_k x_{ik})$$

Substituting for $1 - p(\boldsymbol{x}_i|\boldsymbol{\beta}))$ and simplifying yields

$$l(\boldsymbol{\beta}) = \sum_{i=1}^{n} -ln \, (1 + exp(\beta_0 + \ldots + \beta_k x_{ik})) + \sum_{i=1}^{n} y_i \, (\beta_0 + \ldots + \beta_k x_{ik})$$

Gradient Descent

Now that we have an expression for the log-likelihood, we want to select the parameter estimates that maximize it. Unfortunately, it is not possible to follow our approach from the linear regression model (take partial derivatives, set equal to zero, and find the critical point), as there is no closed-form solution. Instead, we can use a numerical approach to find an approximate solution. The convention in optimization is to solve a minimization problem, and maximizing the log-likelihood will be reframed as minimizing the negative log-likelihood. There are many algorithms to accomplish this task, and we will use gradient descent as it is widespread in machine learning and less complex than many alternatives.

Gradient descent is an iterative method that simultaneously updates the parameters according to this equation:

$$\beta^{(t+1)} = \beta^{(t)} - \alpha \frac{\partial l(\beta^{(t)})}{\partial \beta^{(t)}}$$

where

- $\beta^{(t)}$ is the vector of parameter estimates at iteration t

- α is the learning rate or step size

- $\partial l(\beta^{(t)})/\partial \beta^{(t)}$ is the gradient of the log-likelihood function

A useful analogy for understanding gradient descent is a hiker finding the lowest point in the mountains. Imagine that she moves downhill in the direction of greatest steepness (the gradient) for a given number of paces (the learning rate), and then repeats this process. After a certain number of iterations, she will reach the lowest point (the global minimum) or a point which is lowest in the nearby area (the local minimum).

The algorithm begins by initializing the parameter estimates as $\beta^{(0)}$. The parameter update moves in the direction of the largest change, which is determined by the gradient. For a small enough step size, moving in the direction of the gradient will provide an update to the parameter estimates which brings them closer to their optimal values. A larger step size can lead to faster convergence, but it might miss the minimum. When optimality is reached, the algorithm is said to converge. In practice, it may not be possible to reach optimality (due to a bad starting guess), or it may take a very long time (due to a small α). Stopping criteria are used to prevent the algorithm from running indefinitely: (1) a maximum number of iterations is provided, and (2) a tolerance is set to determine when a change in the objective function is small enough to stop the algorithm.

We will need a formula for the partial derivative of the log-likelihood with respect to each parameter β_j. This includes taking the derivative of the function $ln\,(1 + exp(\beta_0 + \beta_1 x_{i1} + \ldots + \beta_k x_{ik}))$. Recall that the differentiation rule for the logarithm looks like this:

$$\frac{d}{dx}ln(x) = \frac{1}{x}$$

Here, we have a composite function, so the chain rule can be used. For example,

$$\frac{d}{d\beta}ln(1 + \beta x) = \frac{1}{1 + \beta x} \cdot \frac{d}{d\beta}(1 + \beta x)$$

$$= \frac{x}{1 + \beta x}$$

The function we differentiate also includes an exponential term. As a reminder, the differentiation rule for the exponential looks like this:

$$\frac{d}{dx}e^x = e^x$$

We have a composite function of the exponential, so the chain rule can again be used. For example,

$$\frac{d}{d\beta}e^{\beta x} = e^{\beta x} \cdot \frac{d}{d\beta}(\beta x)$$

$$= xe^{\beta x}$$

We now have all the pieces that we need, and the equation for the log-likelihood is repeated here:

$$l(\beta) = \sum_{i=1}^{n} -ln\left(1 + exp(\beta_0 + \ldots + \beta_k x_{ik})\right) + \sum_{i=1}^{n} y_i\left(\beta_0 + \ldots + \beta_k x_{ik}\right)$$

Applying the differentiation rule for the logarithm with the chain rule yields the partial derivative for parameter j:

$$\frac{\partial l(\beta)}{\partial \beta_j} = -\sum_{i=1}^{n} \frac{exp(\beta_0 + \beta_1 x_{i1} + \ldots + \beta_k x_{ik})}{1 + exp(\beta_0 + \beta_1 x_{i1} + \ldots + \beta_k x_{ik})}x_{ij} + \sum_{i=1}^{n} y_i x_{ij}$$

The term

$$\frac{exp(\beta_0 + \beta_1 x_{i1} + \ldots + \beta_k x_{ik})}{1 + exp(\beta_0 + \beta_1 x_{i1} + \ldots + \beta_k x_{ik})}$$

is equivalent to $p(x_i|\beta)$ (this will be left as an exercise), and so we can simplify the equation to:

$$\frac{\partial l(\boldsymbol{\beta})}{\partial \beta_j} = -\sum_{i=1}^{n} p(\boldsymbol{x}_i|\boldsymbol{\beta})x_{ij} + \sum_{i=1}^{n} y_i x_{ij}$$

$$= \sum_{i=1}^{n} (y_i - p(\boldsymbol{x}_i|\boldsymbol{\beta}))x_{ij}$$

Code to run gradient descent is provided below. The essential functions are the sigmoid $p(\boldsymbol{x}_i|\boldsymbol{\beta})$, the log-likelihood, and the gradient descent algorithm.

```python
def sigmoid(t):
    return 1 / ( 1 + np.exp(-t) )



def log_likelihood(X, y, betas):

    # linear combination of predictors
    t = np.dot(X, betas)

    ll = np.sum( y * t - np.log(1 + np.exp(t)) )
    return ll



def gradient_descent_logistic(X, y, alpha=1e-3, n_iter=1000):
    '''
    INPUTS
    X          numpy array with observations on rows,
               predictors on columns
    y          numpy array with labels
    alpha      float, learning rate
    n_iter     int, number of iterations to run

    OUTPUTS
    betas      numpy array of parameter estimates
    ll         list of log-likelihoods
    '''

    # storage for log likelihoods
    ll = []

    # column of 1s for intercept term
    ones = np.ones((X.shape[0], 1))
```

```
# prepend column of 1s to design matrix
X = np.hstack((ones, X))

# initialize parameter estimates
betas = np.zeros(X.shape[1])

for it in np.arange(n_iter):

    # linear combination of predictors
    t = np.dot(X, betas)

    # predicted probability of event for each observation
    p = sigmoid(t)

    # sum over each observation:
    # product of predictors, prediction error y-p
    # this gives the gradient
    # of negative log likelihood
    grad_neg_ll = -np.dot(X.T, y - p)

    # parameter update equation
    betas = betas - alpha * grad_neg_ll

    # for every 1000 iterations,
    # print iteration and append log likelihood
    if it % 1000 == 0:
        print('iter',it)
        ll.append(log_likelihood(X, y, betas))

return betas, ll
```

Let's walk through the gradient descent algorithm. For understanding the calculations and doing matrix algebra, it helps to look at the object shapes. We will take a particular dataset, which we will study in more detail in the next section. The predictor data x_tr has 341 observations and 2 predictors for shape (341,2). The labels y_tr are in a vector of length 341. An empty list is created for storing the log-likelihood. A column of 1s is created for the intercept term. It is placed as the first column in the design matrix using hstack(); the shape of x_tr is now (341,3). The parameter estimates (betas) are initialized to a column vector of zeros with length 3. It will hold the intercept and two predictors. If we check the shape of the betas, it will be reported as (3,).

The code then enters the loop, which will iteratively update the betas. The linear combination of predictors is calculated by computing a matrix product between the predictor values with shape (341,3) and the betas with shape (3,). As the inner dimensions of the product match, this is a valid operation which will produce a column vector with length 341. Next, the sigmoid function computes the probability of the event for each of the 341 observations. It can do this without a loop, as it is a vectorized function.

Next, we need to compute the gradient of the negative log-likelihood, which provides the update direction of the parameter vector. Mathematically, this has form

$$\sum_{i=1}^{n} -(y_i - p(\boldsymbol{x}_i|\boldsymbol{\beta}))x_{ij}$$

This can be computed compactly with the matrix operation

```
-np.dot(X.T, y - p)
```

Let's break this down. Both y and p have the same shape, and we can compute their difference. The term $y - p$ represents the prediction errors for each observation. For each parameter j, we need to compute the dot product between the prediction errors and the values of the jth predictor; this is summing over the observations. The end result is a column vector of length 3, where element j represents a partial derivative with respect to parameter j.

Now that the gradient is computed, we update the parameter estimates taking a step of size α in the direction of the negative gradient. This is expected to minimize the negative log-likelihood (or maximize the log-likelihood). For every 1000 iterations, we report the iteration number (the progress) and store the log-likelihood. If we plot the log-likelihood versus the iterations, we should see it converging to a value, or limit. For an appropriate α and number of iterations, the vector of parameter estimates should converge to the true values.

In the next section, we will consider a numerical example. We will first fit a model using `sklearn`, and then we will call `gradient_descent_logistic()` to show that the parameter estimates converge.

14.2   Detecting Breast Cancer

Next, we will fit a small logistic regression model to predict if a tumor is benign or malignant. This model is reasonable as the target variable can take two possible values. The dataset [43] was sourced from the UCI repo:

```
https://archive.ics.uci.edu/ml/datasets/
breast+cancer+wisconsin (diagnostic)
```

The notebook can be found in the course repo here:

```
semester2/week_03_04_classification_and_logistic_regression/
logistic_regression_w_breast_cancer_data.ipynb
```

The target variable is called `diagnosis` and it takes value 'M' for malignant and 'B' for benign. Each patient has a unique identifier saved in the `id` column, and the columns `f1-f30` are cell measurements that can be used as predictors. We begin by importing modules and reading in the data:

```
import numpy as np
import pandas as pd
from sklearn.linear_model import LogisticRegression
from sklearn.model_selection import train_test_split

datapath = '../datasets/wdbc.csv'
df = pd.read_csv(datapath)
```

We will need to code the target values as 1 for the event 'M' and 0 for the non-event 'B.' The predictors are selected, and the dataset is split into a training set and test set for performance measurement.

```
# code the target
df['target'] = df['diagnosis'].apply(lambda x: 1 if x == 'M'
                                               else 0)

# select the data for the model, saving as numpy arrays
X = df[['f1','f2']].values
y = df['target'].values

# place 60% of data in training set, 40% in test set
x_tr, x_te, y_tr, y_te = train_test_split(X, y,
                                 train_size = 0.6,
                                 random_state=314)
```

Let's look at the first five rows of training data:

```
print('x_tr: \n', x_tr[:5,:])
print('')
print('y_tr: \n', y_tr[:5])
```

OUTPUT:

```
x_tr:
 [[19.21 18.57]
 [19.59 25.  ]
 [10.29 27.61]
 [13.85 19.6 ]
 [12.47 18.6 ]]
```

```
y_tr:
 [1 1 0 0 0]
```

Next, we train the model on the training data. There are several parameters that can be set, such as the maximum number of iterations, but the default settings will be used here. To turn off the penalty for large parameter values, we include **penalty='none'**.

```
model = LogisticRegression(penalty='none').fit(x_tr, y_tr)
```

We can produce the predicted probability of each class by calling the **predict_proba()** function. Here are probabilities for the first five subjects in the training set:

```
model.predict_proba(x_tr)[:5,:]
```

```
OUTPUT
array([[0.00893677, 0.99106323],
       [0.00135816, 0.99864184],
       [0.95568155, 0.04431845],
       [0.7259081 , 0.2740919 ],
       [0.93858145, 0.06141855]])
```

For each row, the values are the probabilities that the tumor is benign (negative class) and malignant, respectively. For example, for the first subject, the probability of a benign cell is 0.00893677, and the probability of a malignant cell is 0.99106323. Since the malignant probability is greater than the default threshold of 0.5, the predicted cell type is 1 (malignant).

Suppose we want to change the threshold and predict malignancy if the probability of the positive label is greater than 0.85. The threshold adjustment can be done like this:

```
model.predict_proba(x_tr)[:,1] > 0.85
```

This will compute the predicted probabilities for each subject and compare the positive-label probabilities against the threshold for each subject. Since we now need to be more confident in predicting malignancy, we can expect the precision to be higher (fewer false positives) but the recall to be lower (more false negatives).

The **sklearn** package allows for easily training the model, but it abstracts away the mathematics. To dive deeper, we will make the connection between what the code is doing and how the sigmoid function works. We can extract the parameter estimates for the intercept and slopes, and apply the sigmoid function to calculate the probability of malignancy for a sample patient.

```
# parameter estimates
b0 = model.intercept_
b1 = model.coef_[0][0]
b2 = model.coef_[0][1]

# data from first subject in training set
x1 = x_tr[0][0] # first row, first column
x2 = x_tr[0][1] # first row, second column

# sigmoid
1 / ( 1 + np.exp(-(b0 + b1 * x1 + b2 * x2) ))

OUTPUT:
array([0.99106323])
```

This value matches the probability from the `predict_proba()` function.

14.2.1 Interpreting the Parameter Estimates

To understand which predictors are statistically significant, we return to the statsmodels package. The `Logit()` object supports logistic regression, and the code and results are shown below.

```
# import statsmodels
import statsmodels.api as sm

# append column of 1s for intercept term
Xtr = sm.add_constant(x_tr)

# fit the logistic regression model
results = sm.Logit(y_tr, Xtr).fit()

# print the model summary
print(results.summary())
```

Table 14.1 shows the parameter estimate table from the model summary. The p-values and confidence intervals indicate that the intercept and predictors are significant. The positive sign on predictors x1 and x2 indicates that they both increase the probability of malignancy. Next, we will learn how to compute the magnitude of the increase.

Since the model relates the log odds to the linear combination of predictors, the parameter interpretation is different from the linear regression case. We

TABLE 14.1: Breast Cancer Parameter Estimate Table

	coef	std err	z	$P > |z|$	[0.025	0.975]
const	−20.7540	2.381	−8.716	0.000	−25.421	−16.087
x1	1.1042	0.134	8.215	0.000	0.841	1.368
x2	0.2289	0.049	4.655	0.000	0.133	0.325

can think about two subjects A and B with identical data except for a single predictor where the value differs by 1 unit. Suppose the data looks like this:

subject	x1	x2
A	v+1	w
B	v	w

In this case, the x1 variable differs by one unit between the subjects. This can be done with any predictor to yield an analogous interpretation. We will now write a formula that compares these two subjects with their data. Recall the formula for the odds of the event, with the linear combination of predictors substituted:

$$\frac{p(\mathbf{x})}{1 - p(\mathbf{x})} = exp(\beta_0 + \beta_1 x_{i1} + \ldots + \beta_k x_{ik})$$

Denote subject A data as $\mathbf{x}_A$ and subject B data as $\mathbf{x}_B$. Next, we form the ratio of the odds, called the *odds ratio* (abbreviated OR) for the two subjects like this:

$$OR = \frac{p(\mathbf{x}_A)/(1 - p(\mathbf{x}_A))}{p(\mathbf{x}_B)/(1 - p(\mathbf{x}_B))}$$

Now we will form an equation for the odds ratio, using the linear combination of predictors for each subject and plugging in the data:

$$\frac{p(\mathbf{x}_A)/(1 - p(\mathbf{x}_A))}{p(\mathbf{x}_B)/(1 - p(\mathbf{x}_B))} = \frac{exp(\beta_0 + \beta_1(v + 1) + \beta_2 w)}{exp(\beta_0 + \beta_1 v + \beta_2 w)}$$

We can use the property of exponentials:

$$exp(a)/exp(b) = exp(a - b)$$

This will simplify the ratio greatly, since for example

$$exp(\beta_0)/exp(\beta_0) = exp(\beta_0 - \beta_0)$$
$$= exp(0)$$
$$= 1$$

More generally, the same term in the numerator and denominator yields a multiplicative factor of 1.

Also recall the following property of exponentials:

$$exp(a + b) = exp(a)exp(b)$$

This allows for writing simple factors in the numerator and denominator, which cancel. Returning to the odds ratio, the equation becomes

$$
\begin{aligned}
\frac{p(\mathbf{x}_A)/(1 - p(\mathbf{x}_A))}{p(\mathbf{x}_B)/(1 - p(\mathbf{x}_B))} &= \frac{exp(\beta_0 + \beta_1(v + 1) + \beta_2 w)}{exp(\beta_0 + \beta_1 v + \beta_2 w)} \\
&= \frac{exp(\beta_0) \times exp(\beta_1(v + 1)) \times exp(\beta_2 w)}{exp(\beta_0) \times exp(\beta_1 v) \times exp(\beta_2 w)} \\
&= exp(\beta_1[(v + 1) - v]) \\
&= exp(\beta_1[1]) \\
&= exp(\beta_1)
\end{aligned}
$$

The odds ratio has a very simple form and interpretation: increasing the value of predictor j by one unit will multiply the odds of the event by the factor $exp(\beta_j)$. Returning to the parameter estimate table, the coefficient on `x1` is 1.1042. Increasing `x1` by one unit will multiply the odds of malignancy by a factor of $e^{1.1042} = 3.017$.

To understand from the parameter estimate if the odds will increase, decrease, or remain unchanged, we can consider the case where the value is zero. In this case, $e^0 = 1$, which means that the odds are multiplied by a factor of one (it is unchanged). A positive parameter estimate will increase the odds of the event, while a negative parameter estimate will decrease the odds of the event.

14.2.2 Revisiting Gradient Descent

From fitting the model with packages, we found the parameter estimates to be

$$[-20.7540, \ 1.1042, \ 0.2289]$$

Let's see if gradient descent produces a similar result. We will set a relatively small alpha and a large number of iterations:

```
betas, ll = gradient_descent_logistic(x_tr,
                                      y_tr,
                                      alpha=1e-4,
                                      n_iter=2e6)
```

The betas from gradient descent are

$$[-20.75395723, \quad 1.1041736, \quad 0.22893847]$$

Rounding to four decimal places, the results match. Figure 14.2 plots the log-likelihood versus the number of iterations. The curve increases to a stable value relatively quickly. Hopefully, this illustration provided a greater understanding of how the parameter estimates can be computed. There is a wide range of models in machine learning, and the parameters are often estimated using some form of gradient descent.

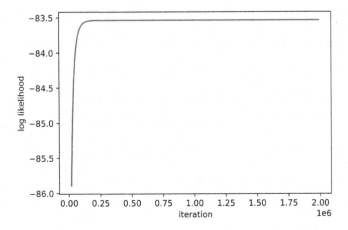

FIGURE 14.2: Convergence of Gradient Descent

14.2.3   Model Performance Evaluation

In this section, we will compare predictions on the test set to the actual labels. When producing predicted labels from predicted probabilities, a threshold needs to be applied. The threshold in `sklearn` is 0.50 by default. A different threshold can sometimes give better results, and we will consider this next. The metrics we will discuss are accuracy, recall, precision, and F1 score. These were introduced in Chapter 11. There are several ways to calculate the metrics: according to the definition, with an individual function from `sklearn` such as `recall()`, or with the function `classification_report()` which computes metrics in batch. Let's first run `classification_report()` and review the output. The function takes the actual labels and predicted labels, in this order.

```
print(classification_report(y_te, model.predict(x_te)))
            precision    recall  f1-score   support
```

0	0.88	0.93	0.91	151
1	0.85	0.75	0.80	77
accuracy			0.87	228
macro avg	0.87	0.84	0.85	228
weighted avg	0.87	0.87	0.87	228

The first section in the report shows metrics for the cases with a negative label (0) and a positive label (1). Focusing on row (1), the precision indicates that 85% of the predicted positives were truly positive. The recall indicates that 75% of the true positives were predicted as positive. The F1 score is the harmonic mean of the precision and recall, valued at 0.80 or 80%. The *support* indicates there were 77 positive-labeled cases. The calculations for row (0) are analogous, and they represent the classification ability for the negative-labeled cases. Given the F1 score of 0.91, we learn that the model does better on negative labels than positive labels. From the second section in the table, we see that the accuracy, or fraction of correct predictions, was 0.87, or 87%. For data where the number of positive labels is different from the number of negative labels, the accuracy is not very useful.

In practice, it can be helpful to set a range of thresholds and compute recall and precision for each. The threshold giving the highest F1 score, for example, might then be selected. Next, we report metrics for a sample of thresholds:

```
# probabilities of positive class from test set
pred_pos = model.predict_proba(x_te)[:,1]

# calculate metrics over range of thresholds
pre, re, th = precision_recall_curve(y_te,  pred_pos)

print('threshold:', th[100:105])
print('')
print('precision:', pre[100:105])
print('')
print('recall:   ', re[100:105])
```

OUTPUT:

```
threshold: [0.5232687 0.5311383 0.5393744 0.5536759 0.563716]

precision: [0.8769231 0.890625   0.904762   0.9032258 0.9016393]

recall:    [0.7402597 0.7402597 0.7402597 0.7272727 0.7142857]
```

For threshold 0.523, the precision and recall are approximately 0.87 and 0.74, respectively. Moving the threshold up to 0.564 (the last shown threshold) increases the precision and decreases the recall. This makes sense, as a higher threshold makes it more stringent to predict a positive label. This reduces the number of false positives, which increases precision. At the same time, there are fewer predicted positives, and this increases the number of false negatives (cases which were actually positive but predicted as negative). The F1 scores for the lower and higher thresholds are 0.8028 and 0.7971, respectively, which suggests that the lower threshold is the better of the two.

A different model will likely produce different metrics. While the threshold may be changed to give better results, it is always preferable to build the best model possible, and then try different thresholds to tune it. The model and optimal threshold can then be saved to predict outcomes on new data.

14.3   Chapter Summary

Logistic regression can be a useful supervised learning approach when the target variable is binary. It assumes that a linear combination of predictors is linearly related to the log odds of an event. To estimate the parameters, we worked with the log-likelihood of the data and applied the gradient descent algorithm. Gradient descent works by iteratively moving the parameter estimates in a direction which is expected to be optimal. Many machine learning models apply variants of gradient descent, including deep neural networks.

We considered a dataset of tumors which were either benign or malignant. This was a good use case for logistic regression, and we fit the model using `sklearn`. We also verified the parameter estimates using gradient descent, and they matched given our learning rate and number of iterations. We measured model performance using binary classification metrics. As the model outputs probabilities, the predicted label is determined by applying a threshold to the probabilities. We learned that varying the threshold will produce different performance metrics.

In this chapter and the previous chapter, we studied models that are useful when the data is labeled. Oftentimes, data is not labeled, and unsupervised methods will be valuable. In the next chapter, we will study an algorithm that is useful for exploring data groupings and detecting outliers without the benefit of labels.

14.4   Exercises

Exercises with solutions are marked with Ⓢ. Solutions can be found in the `book_materials` folder of the course repo at this location:

https://github.com/PredictioNN/intro_data_science_course/

1. ⓢ Explain why the direct use of a linear combination of predictors is not appropriate for the logistic regression model.

2. What is the value of the sigmoid function $\sigma(t)$ for $t = 0$?

3. ⓢ If the odds of an event are $1/3$, what is the probability of the event?

4. Explain what the likelihood function represents.

5. ⓢ True or False: The parameters that maximize the likelihood will also maximize the log-likelihood.

6. In a logistic regression model, what is the assumed relationship between the log odds of an event and the linear combination of predictors?

7. Is it possible to estimate the parameters of a logistic regression model by finding the critical point? Explain your answer.

8. In gradient descent, what does the step size represent?

9. ⓢ In gradient descent, does it always make sense to use a large step size? Explain your answer.

10. ⓢ In computing the gradient of the log-likelihood, we made a substitution for the event probability. Show that the representations below are equivalent.

Expression I

$$\frac{exp(\beta_0 + \beta_1 x_{i1} + \ldots + \beta_k x_{ik})}{1 + exp(\beta_0 + \beta_1 x_{i1} + \ldots + \beta_k x_{ik})}$$

Expression II

$$\frac{1}{1 + exp(-(\beta_0 + \beta_1 x_{i1} + \ldots + \beta_k x_{ik}))}$$

11. What will be the result of running the following code:

```
import numpy as np

a = np.array((1,2,3))
b = np.array([(1,2,3),(1,2,3)])

np.hstack((a,b))
```

12. In the function `gradient_descent_logistic()`, explain what this line is doing:

    ```
    betas = betas - alpha * grad_neg_ll
    ```

13. Ⓢ True or False: When the logistic regression parameter estimates are computed with gradient descent, we should expect them to be very different from estimates derived by a statistical package like `statsmodels`.

14. You have fit a logistic regression model with a single binary predictor. The coefficient estimate for the predictor has value zero. Consider one subject with a value of 0 for the predictor, and another subject with a value of 1 for the predictor. How will the odds of the event differ between the subjects?

15. Ⓢ Using the `sklearn` module, fit a logistic regression model to this dataset: $\{(3,0),(3.2,0),(4,1)\}$. Show your code and print the intercept and slope coefficient estimates. For a predictor value of 6, what is the predicted target?

16. Ⓢ Explain the difference between the `sklearn` functions `predict()` and `predict_proba()`. Specifically, when would it make sense to use each of them?

17. A logistic regression model produced 100 true positives and a precision of 0.2. How many false positives were there?

18. Ⓢ A logistic regression model was applied to a live dataset. The resulting precision was deemed too low by product leadership. What can be done to the probability threshold to produce a higher precision?

15

Clustering with K-Means

We don't always have the luxury of labeled data for supervised learning. A more common use case is that a dataset $\{\mathbf{x}_i\}_{i=1}^n$ is not labeled, but we want to identify substructure such as groupings. Clustering techniques are also helpful in finding outliers; such points may be located in isolated clusters or small groups. We will begin this chapter by studying concepts essential to clustering. Then we will learn about k-means clustering and work out a small example. Lastly, we will apply k-means to a nutrition dataset of roughly 9000 foods.

15.1   Clustering Concepts

Let's start with a very small example to illustrate the ideas. Suppose that a teacher wishes to group her students by some measure of assessment performance. The assessments may consist of quizzes or exams, for example. After the students are grouped, they can be given more targeted instruction. What is the best way to implement this grouping? We face some questions to be answered early on, including:

- Which students should be used in the grouping calculations? We might include in the analysis all current students in a specific class, or we might use both current and historical students from a course.

- Which variables should be included? We might include one assessment at a time, all assessments given in a particular course, or only the assessments given before the final exam. For example, if we want a system that can identify students needing support on recent learning, then the clustering could be based on the last exam.

- How do we measure the quality of the groups? To understand the quality of the groups, we might try to correlate the groupings with another variable. It is also important to bear in mind that there is a limit to the number of groups that can be supported by the teacher and teaching assistants.

There is no single, prescriptive answer to these questions. Instead, the right choices will depend on how the clustering information will be used. It will also take some iterating to arrive at a satisfactory answer.

Let's narrow the focus to three students who have taken two exams. Their scores are as follows:

student	exam1	exam2
A	70	82
B	75	80
C	95	90

What is a reasonable way to group these students based on their exam scores? First, we would probably agree that a group, or *cluster*, should consist of students who are more similar to each other. If two students are different, then they should be in different clusters. To measure similarity, we can represent each student as a point based on data. We can then measure the distance between points using a distance metric. There are many possible distance metrics, and the particular problem might define the metric for us. In this example, we will use *Euclidean distance*, which is a straight-line distance. Figure 15.1 provides an illustration with students A and B as points.

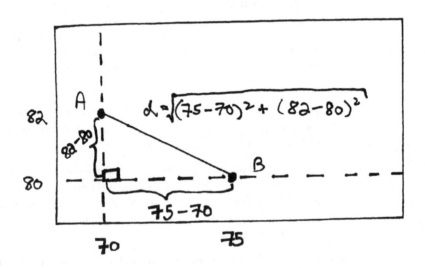

FIGURE 15.1: Calculating Euclidean Distance

Notice where the horizontal and vertical lines cross, as it shows three things:

- The intersection forms a right angle

- Student A scored a 70 on exam 1

- Student B scored an 80 on exam 2

The distance between A and B will be the hypotenuse of the formed right triangle. The horizontal side of the triangle represents the difference in scores for exam 1, 75 - 70, while the vertical side represents the difference in scores for exam 2, 82 - 80. The distance may then be calculated as $\sqrt{(75 - 70)^2 + (82 - 80)^2} = \sqrt{29} \approx 5.39$ according to the Pythagorean theorem (which you might recall as $c = \sqrt{a^2 + b^2}$).

Euclidean distance can be stated as a formula for two vectors $\mathbf{x}$ and $\mathbf{y}$, each with length p:

$$d(\mathbf{x}, \mathbf{y}) = \sqrt{\sum_{i=1}^{p}(x_i - y_i)^2} \tag{15.1}$$

This is simply an extension of our example to the p-dimensional case. The distances between the other students could be calculated in a similar fashion. Notice that one method to do grouping involves the calculation of all pairwise distances: $d(A, B), d(A, C), d(B, C)$. For a large number of datapoints, this can be a large calculation. One simplification, which we will see later with the k-means algorithm, is to stop short of taking the square root. Instead, we can calculate squared distances d^2, because ranking based on d^2 will be identical to ranking based on d. This is because the square root function is a monotonically increasing function.

Another consideration is the number of groups k to be formed. A single group won't illuminate similarities or differences in the students, so we should form multiple groups. Three groups won't be much help either, as this places each student on their own. In this case, it makes sense to use two groups. In the general case of n objects, we should require that k is greater than one and less than n. There are different ways of measuring the quality of grouping, and we will study a specific metric shortly. Conceptually, points within a group should be "close together" and points in different groups should be "far apart." The closeness can be measured by sums of squared distances between points.

In the next section, we will discuss the popular k-means clustering algorithm. The algorithm can scale to massive numbers of observations and variables, and it requires only a single parameter. We have already introduced many of the important concepts used in k-means, so let's dive deeper.

15.2   K-Means

The k-means algorithm divides n objects, or points, into k clusters by placing
each point in the cluster with the nearest *centroid*. Figure 15.2 shows k-means
clustering for the three-student example. The clustering uses two clusters, and
the × marks show the centroids. To understand how the centroids are decided,
we'll need to study the steps of the algorithm, which we do next.

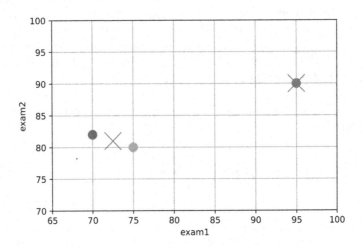

FIGURE 15.2: Data Points and K-Means Centroids

Given the number of clusters k, there are different methods for initializing the
centroids. A common approach is to randomly select k of the points and make
them centroids.

Each point is then assigned to the closest centroid based on Euclidean distance.
This is the *assignment step*.

Next, each centroid is recalculated by selecting the points assigned to it and
computing the average value of each variable. For p variables, each centroid
will be a p-dimensional vector where the ith element is the average value of
the ith variable. If a cluster consists of points $(1, 2)$ and $(3, 4)$, for example,
the centroid would be $((1 + 3)/2, (2 + 4)/2) = (2, 3)$. This step is called the
update step.

The pair of steps (assignment, update) is repeated until there are no assignment changes or the algorithm has run the allowed number of iterations.

The k-means algorithm attempts to minimize within-cluster sum of squared
errors (WSSE). This would satisfy the desire for points in the same cluster

to be "close together." Unfortunately, k-means is not guaranteed to attain its objective, and the result depends on the initial centroids. For this reason, the algorithm should be run multiple times and the best result should be used. This is implemented in software, and we will use `sklearn`.

K-Means Assumptions and Limitations

K-means has several assumptions and limitations that should be understood:

1) Since the algorithm uses Euclidean distance, the clustering pattern of the data will be important. Specifically, it will do well when the clusters are spherically shaped. It won't do well if there are concentric rings of data or complex shapes; an alternative method like DBSCAN will work better for these cases.

2) As distance measurements are used in clustering, it is important to ensure that the variables are on the same scale. Normalization or scaling can be used for this purpose.

3) The variables need to be numeric. Categorical variables are typically one-hot encoded.

4) K-means will always give an answer. It may take a large amount of time to evaluate the results, iterate, and understand if the clusters are useful.

5) The results can be highly dependent on the variables and observations used. For example, including or dropping a variable can produce different clustering results. This should be explored as part of testing.

15.2.1   K-Means Hand Calculation

To clearly understand an algorithm, a hand calculation can be very helpful. We can work out k-means by hand for our small example. Note that because the number of points and variables is so small, we could try each initialization, select the best one, and be guaranteed to reach optimality. For our purposes, it will be sufficient to illustrate a single initialization. A large-scale case would not allow for exhaustive search (trying each initialization) due to the massive number of configurations.

As a reminder, here is the student data:

student	exam1	exam2
A	70	82
B	75	80
C	95	90

Initialization

Let $k = 2$. Initialize the process by placing centroid 0 at A and centroid 1 at C. These points are used as the initial averages.

centroid	exam1_avg	exam2_avg
0	70	82
1	95	90

Iteration 1: Assign

Based on the centroids, we need to assign the points to their closest centroid. To do this, we first need to measure the distance of each point P to each centroid as follows:

student	dist(P,centroid0)	dist(P,centroid1)
A	0	26.2
B	5.4	22.4
C	26.2	0

For example, $d(A, centroid1) = \sqrt{(95 - 70)^2 + (90 - 82)^2} = 26.2$. Now we can make assignments:

student	exam1	exam2	centroid
A	70	82	0
B	75	80	0
C	95	90	1

Students A and C are at centroids, and so their assignments are unchanged. Student B is much closer to cluster 0 (distance=5.4) than cluster 1 (distance=22.4).

Iteration 1: Update

Now we update the centroids by selecting students in each cluster and computing the averages of each exam. For centroid 0, we average exams for students A and B. Only student C is assigned to centroid 1. Here are the updated centroids:

centroid	exam1_avg	exam2_avg
0	72.5	81
1	95	90

Iteration 2: Assign

Given the updated centroids, we need to recompute distances between the points and centroids. Here are the new values:

student	dist(P,centroid0)	dist(P,centroid1)
A	2.7	26.2
B	2.7	22.4
C	24.2	0

Since centroid 1 hasn't moved, its distances from students A and B are unchanged; this means the points' assignments are unchanged. Student C is already at a centroid, thus its assignment won't change.

Iteration 2: Update
Since the assignments haven't changed, the centroids won't change and the algorithm stops. We now have final cluster assignments:

student	centroid
A	0
B	0
C	1

Each centroid provides predicted values for the points in its cluster. Student A earned a 70 on exam 1 and an 82 on exam 2. The student is assigned to cluster 0, which has a centroid with a predicted exam 1 score of 72.5 and a predicted exam 2 score of 81. Likewise, students B and C each have two exam scores and two predicted scores. Similar to the case of regression, we can't measure accuracy by summing deviations between actual values and predictions, since cancellations will occur. Instead, we compute sums of squared deviations.

This brings us back to the objective of k-means, which is to minimize the within-cluster sum of squared errors. The calculation of WSSE is organized in the following table, where columns 2 and 3 show the squared differences between student exam scores and predicted scores. The WSSE is equal to the sum over all cells in this table, which is 14.5. If we tried other initializations, we would use the results (centroids, cluster assignments, etc.) which produced the lowest WSSE.

student	exam1_sq_error	exam2_sq_error
A	$(70 - 72.5)^2$	$(82 - 81)^2$
B	$(75 - 72.5)^2$	$(80 - 81)^2$
C	$(95 - 95)^2$	$(90 - 90)^2$

15.2.2   Performance Evaluation

The k-means algorithm attempts to minimize the within-cluster sum of squares for a given number of clusters k. The question remains of how to select the best k. Selecting the k that minimizes WSSE won't work, as the metric can be minimized by setting $k = n$. In this case, each point is its own cluster and WSSE=0, but this tells us nothing about substructure in the data. Something else is needed.

A common approach is to plot the WSSE for a range of k as in Figure 15.3. As k increases, WSSE will generally decrease. There is sometimes a sharp drop and then a flatting of WSSE, which forms an "elbow" pattern. The k corresponding to this elbow is a sensible value for use. For this particular graph, the elbow is not particularly clear, but a k of perhaps 3 or 4 is sensible. Increasing k beyond the underlying number of clusters will break apart points that belong together, yielding little change in WSSE.

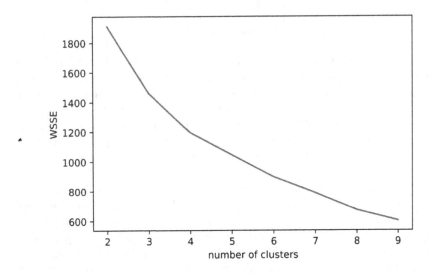

FIGURE 15.3: WSSE versus Number of Clusters

Another more quantitative approach for selecting k is to compute the *silhouette score*. The silhouette score measures the *cohesion* of each data point within its cluster when compared to other clusters. The metric falls in the range $[-1, 1]$ where a higher value indicates better cohesion. Given this, the metric can be used to compare different k and pick the best value. We will limit attention to the case $k > 1$.

Given the assignment of points to clusters, the silhouette score is calculated as follows:

1) For each data point i, calculate its mean distance from all other points in its cluster, calling it $a(i)$. This measures how well the point is aligned in its cluster.

2) For each data point i, compute its mean distance from all of the points in a different cluster. Repeat this mean calculation for each of the different clusters and identify the cluster with smallest mean. This is the *neighboring cluster* of point i, since it is the next-best cluster. The mean distance to points in the neighboring cluster is denoted $b(i)$.

3) For each data point i, calculate its *silhouette* as

$$s(i) = \frac{b(i) - a(i)}{max\{a(i), b(i)\}}$$

A large difference $b(i) - a(i)$ indicates a good fit between point i and its cluster. In the extreme case where each point in a cluster is identical, $a(i) = 0$ for each of these points. The denominator normalizes values to fall in the range $[-1,1]$. For the extreme case mentioned, the numerator would be $b(i) - a(i) = b(i)$ and the denominator would be $max\{a(i), b(i)\} = b(i)$. The silhouettes are then $s(i) = b(i)/b(i) = 1$.

4) The silhouette score is calculated as the mean $s(i)$ over all i. For the case where each point in a cluster is identical, $s(i) = 1$ for all i, and the silhouette score will be 1.

The silhouette score can be calculated in Python using `sklearn` and we will see this illustrated in the next section.

15.3 Clustering Foods by Nutritional Value

Next, we will apply k-means clustering to a nutrition dataset consisting of roughly 9000 common foods and products. There is a wide variety of foods including pecans, teff (a grain), orange sherbet, and eggplant. The data is freely available here:

```
https://www.kaggle.com/datasets/trolukovich/
nutritional-values-for-common-foods-and-products?
resource=download
```

The Jupyter notebook for this demo can be found in the course repo:

```
/book_materials/kmeans_demo.ipynb
```

In the interest of focusing on the new material, we won't comprehensively process this data and iterate heavily on the results. Instead, we will use a subset of foods and nutrition variables. We begin by importing the necessary modules and setting the data path:

```
# import modules
import numpy as np
import matplotlib.pyplot as plt
import pandas as pd
import re
import seaborn as sns
from sklearn.preprocessing import StandardScaler
from sklearn.cluster import KMeans
from sklearn.metrics import silhouette_score

# set path to dataset
data_path = '/path_to_data/nutrition.csv'
```

The `re` package provides functionality for *regular expressions*, which we will use for removing strings from the dataset. In particular, nutritional values appear with their units – such as "10 mg" – and we will need to retain only the numeric portion of the data. We also import functions from `sklearn` to fit and evaluate k-means clustering.

Next we read in the CSV file and specify a list of variables. We will use a subset of the variables to quickly show some results.

```
# import the data
df = pd.read_csv(data_path)

# specify variables
vars = ['calories', 'total_fat', 'saturated_fat', 'cholesterol',
        'sodium', 'choline', 'folate', 'folic_acid', 'niacin',
        'pantothenic_acid', 'riboflavin', 'thiamin',
        'vitamin_a', 'vitamin_a_rae', 'carotene_alpha']
```

It is always important to look at the data and perform exploratory analysis. For this data, there are columns with identical values, such as `serving_size`. This can be detected statistically by computing the standard deviation of each of the variables. If the standard deviation of a variable is zero, then it is constant across records and it should be dropped from further analysis.

Next, we will select only the columns of interest. This consists of the variables and the `name` column, which holds the food names. Note that lists can be added as `list1` + `list2`.

```
df = df[['name'] + vars]
```

A thorough analysis might individually impute missing values for each variable. Outliers should also be treated carefully. Here, we drop any row which has a missing value, and outliers are not treated.

```
# drop rows with any missing values
df = df.dropna(axis=0)
```

The `calories` variable contains integers, while the other variables contain strings due to the inclusion of units such as "10.5mg" or "10.5 mg." Note that in a proper database, units would be in a separate field from values, but data in the wild may have all sorts of complications. Our strategy will be to loop over each variable, determine which are strings, retain only the numeric portions, and cast the values to floats. Then "10.5mg" or "10.5 mg" will become 10.5. The regular expression pattern, or regex, that we will use is this: [.|\d]+. Regexes can be cryptic to the uninitiated, but with exposure and practice comes skill. The plus sign (+) denotes one or more occurrences, and it allows for retaining all dots (.) and digits (\d). For small tasks, simply searching for a regex should give the right form. An excellent tool for building, testing, and debugging regexes is https://regex101.com/

We loop over each variable to prepare it for modeling:

```
for var in vars:

    # clean the column if it doesn't hold integers
    if not isinstance(df[var].values[0], np.int64):

        # use regex to retain numeric part of data
        df[var] = df[var].str.extract(r'([.|\d]+)')

        # convert strings to floats
        df[var] = df[var].values.astype(float)
```

After this completes, all of the predictor data will be numeric. Table 15.1 shows a sample of the rows and columns.

TABLE 15.1: Sample of Cleaned Nutrition Data

name	calories	total_fat	saturated_fat	cholesterol
Nuts, pecans	691	72.0	6.2	0.0
Teff, uncooked	367	2.4	0.4	0.0
Sherbet, orange	144	2.0	1.2	1.0
Cauliflower, raw	25	0.3	0.1	0.0
Taro leaves, raw	42	0.7	0.2	0.0

Observe the different scales of the columns: calories and saturated fat may differ by three orders of magnitude. Since k-means uses Euclidean distance between points, is it essential to scale the data. We apply `StandardScaler()` on the variables as follows:

```
# set up the scaler
scaler = StandardScaler()

# compute the z-scores, excluding name column from dataset
scaled_data = scaler.fit_transform(df.values[:,1:])
```

Note that the scaled data is now in a numpy array. Next, we select a small number of rows for illustration:

```
food_subset = 175
scaled_data_sub = scaled_data[:food_subset]
```

Now that the dataset is numeric and scaled, it is ready for k-means. We will fit k-means using a range of values for k. For each fitting, we compute the silhouette score and save it in a list for subsequent visualization. To improve the chances of the algorithm converging to a global optimum, we set 100 initializations.

```
scores = []
rand = 314

# fit k-means with 2 through 9 clusters
clus_range = np.arange(2,10)
for clus in clus_range:
    print(f'fitting k-means for {clus} clusters')

    # fit k-means
    kmeans = KMeans(n_clusters=clus,
                    random_state=rand,
                    n_init=100).fit(scaled_data_sub)

    # compute silhouette scores
    scores.append(silhouette_score(scaled_data_sub,
                                   kmeans.labels_,
                                   metric='euclidean'))

scores
```

Next we plot the silhouette scores, which are shown in Figure 15.4.

```
plt.bar(clus_range, scores)
plt.xlabel('number of clusters')
plt.ylabel('silhouette score')
```

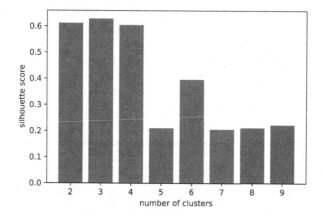

FIGURE 15.4: Silhouette Scores

The clustering with 3 groups edges out the others, and we will use this model. The silhouette score is slightly above 0.6, which indicates cohesion in the clustering. Since we didn't save the earlier models, we refit k-means using $k = 3$. Note the function `np.argmax()` which returns the location holding the maximum score. For example, `np.argmax([1,5,2])`=1 since position 1 holds the maximum of 5 (recall that arrays begin with position 0 in Python).

```
# k with largest score
k_best = clus_range[np.argmax(scores)]

kmeans_final = KMeans(n_clusters=k_best,
                      random_state=rand,
                      n_init=100).fit(scaled_data_sub)
```

Now that we've trained the model, each food has been assigned to a cluster. The number of foods in clusters 0, 1, and 2 is 162, 11, and 2, respectively. The small number of foods in cluster 2 suggests that it may contain outliers; it can be flagged for follow-up.

It will be helpful to review the foods in each cluster, and to observe how their nutritional values differ in aggregate. To do this, we will return to the dataframe, select the appropriate rows and columns, and append the cluster assignments.

```
# for columns: retain name and variables
# for rows: select the subset of foods
df_sub = df[['name']+vars].iloc[:food_subset]

# append the cluster labels
df_sub['cluster'] = labels
```

It will be interesting to look at the means of the variables grouped by cluster. This can be calculated using the `groupby()` and `agg()` functions. The table below shows average calories and fat content for foods in each cluster. For example, if we look at all foods in cluster 0 and compute their average calories, this is 245.111. Cluster 1 contains foods which are higher in calories and total fat, on average, when compared to the other clusters.

```
df_sub.groupby(df_sub.cluster).agg(func=np.mean).round(3)
```

cluster	calories	total_fat	saturated_fat
0	245.111	9.659	2.738
1	721.091	78.273	27.909
2	296.500	8.750	1.550

Figures 15.5 and 15.6 show boxplots of select nutrition variables segmented by cluster. Cluster 1 stands out from the others, with the majority of foods having higher calories and total fat.

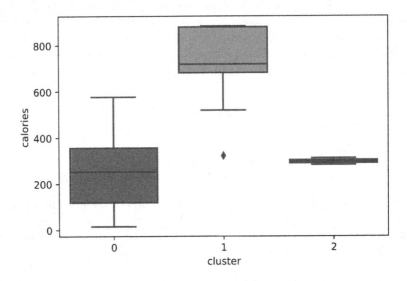

FIGURE 15.5: Boxplot of Calories by Cluster

Lastly, we can print the list of foods in each cluster. In the interest of space, we show up to ten foods per cluster.

```
# display the maximum number of rows
pd.set_option('display.max_rows', None)

# for each cluster, print the food names
```

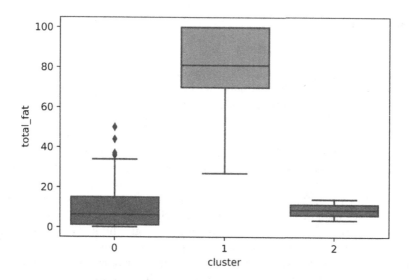

FIGURE 15.6: Boxplot of Total Fat by Cluster

```
for clus in np.arange(k_best):
    print(f'cluster {clus}')
    print(df_sub.name[df_sub.cluster==clus])
    print('')
```

cluster0	cluster1	cluster2
Teff, uncooked	Nuts, pecans	Spices, chili powder
Sherbet, orange	Nuts, dried, pine nuts	Chives, freeze-dried
Cauliflower, raw	Butter oil, anhydrous	
Taro leaves, raw	Oil, soybean lecithin	
Lamb, raw, ground	Egg, fresh, raw, yolk	
Cheese, camembert	Nuts, dried, pilinuts	
Vegetarian fillets	Butter, without salt	
Crackers, rusk toast	Snacks, banana chips	
Chicken, boiled, feet	Oil, corn and canola	
Quail, raw, meat only	Oil, ucuhuba butter	

Cluster 0 is large and diverse. Cluster 1 contains nuts, oils, and butter, which are fattening foods. It is interesting that these foods were grouped. Cluster 2 is very small and may hold outliers. It contains freeze-dried chives, which are roughly 10 times the calories of raw chives. Better clusters might be extracted from this data with more effort.

15.4 Chapter Summary

Unsupervised techniques such as k-means clustering can uncover substructure and outliers. K-means divides n points into k clusters with the objective of minimizing the within-cluster sum of squared errors. Each cluster has a centroid, which is a vector of coordinates where the points in the cluster balance.

K-means begins with initialization of the centroids, and there are several methods for doing this. Following initialization, the algorithm proceeds by assigning each point to its closest centroid based on Euclidean distance. Given the points in the cluster, each centroid is updated to reflect the average value in each dimension. The assignment and update steps are repeated until all iterations are used or the assignments don't change.

Users should be aware of several assumptions and limitations of k-means. The algorithm is not guaranteed to converge to a global optimum. To help find the optimum, k-means should be run a large number of times. Since the algorithm uses Euclidean distance, is it important that variables are on the same scale. The distance metric also has implications for the data patterns that are amenable to k-means: spherical-shaped clusters will yield good results, while complex patterns like concentric circles will not. K-means requires numeric data; categorical variables are typically one-hot encoded.

The only required parameter for K-means is the number of clusters, k. Different techniques are used to determine k, and we investigated the silhouette score. This score measures the fit of each point in its assigned cluster relative to its next-closest cluster. The silhouette score falls in a range of $[-1,1]$ where a higher value is better.

We obtained a large dataset of nutritional values from diverse foods. The dataset required some cleanup, and we removed unwanted strings with a regular expression, or regex. Regexes are useful for finding patterns in data. A k-means cluster analysis of the data found a high-calorie, high-fat cluster. The illustration showed an interesting result with relatively little effort, but a more robust solution would require deeper investigation.

This chapter and earlier chapters focused on machine learning techniques. In the next chapter, we will learn some techniques for making data science projects more reproducible. This will also reduce the time necessary to move a model from development to production.

15.5 Exercises

Exercises with solutions are marked with Ⓢ. Solutions can be found in the `book_materials` folder of the course repo at this location:

https://github.com/PredictioNN/intro_data_science_course/

1. Explain the concept of a centroid.

2. Ⓢ Explain how the k-means algorithm can be used to detect outliers.

3. Ⓢ Provide an example of a data pattern that would not be amenable to k-means clustering.

4. Do you think there are cases where k-means can run indefinitely? Explain your answer.

5. Ⓢ You are deciding on the appropriate value for k when performing k-means clustering. Explain why increasing k until the within-cluster sum of squared errors is minimized is not a good approach.

6. When you rerun k-means, you notice that the results are different. Explain why this might happen.

7. Ⓢ After running k-means with many variables, you notice that the resulting cluster assignments align very closely to one variable in particular. You then realize that you've missed a processing step that led to this variable dominating the others. What is the missing step?

8. Ⓢ To improve the chance of k-means finding the global optimum, the algorithm is run using many different initializations. Why isn't exhaustive search used instead?

9. Select a dataset and use `sklearn` to fit k-means and calculate the silhouette score. Next, verify the silhouette score by calculating it yourself. You might write a Python function to do this.

10. The nutrition example in this chapter used a subset of predictors and light preprocessing to form three clusters. Try some further cleaning and additional predictors to produce a better clustering of the foods. Your results might surface additional clusters with meaningful structure.

16

Elements of Reproducible Data Science

It is a great outcome when a model from the data science team is identified for deployment to production. At the same time, the process of moving the model from development to production takes a lot of work. For organizations that do not have a lot of experience productionizing machine learning models, the steps can seem unclear, onerous, and manual. Concrete, common steps include:

- Sharing code

- Setting up testing and production environments with sufficient software and hardware (e.g., dependencies)

- Testing the model and data

- Insuring that the data pipelines are properly processing the data and feeding the model

- Setting up a secure API endpoint to serve requests

- Collecting and storing artifacts and predictions

- Returning results at the appropriate frequency (e.g., batch or real time)

- Standing up a model monitoring system

- Setting up a pipeline for model retraining

These steps fall in the domain of *MLOps*, which is often handled by members of the engineering team. While a different team owns this process, data scientists will need to actively contribute their knowledge. For example, they will advise on which metrics should be monitored. The better that data scientists understand the path to production, the more they can help. This can allow for a smoother handoff of the model, saving time and effort. Additionally, tight integration between data science and MLOps can help ensure that things are working as expected.

In this chapter, we will explore some of the items listed above. In particular, we will focus on the practices that help make data science products reproducible. The first habit is to maintain code in a collaborative repository such as `GitHub`. The second habit is to develop and include tests with the software. The third practice is to create temporary, reproducible environments with *container* technology. Also note that data science reproducibility pertains to the broadest sense of allowing anyone to easily run code. Examples may include an engineering team, another data scientist, or a group of students running Python in an environment.

Since this is an introductory book on data science, advanced topics like *continuous integration / continuous deployment (CI/CD)* and data versioning are out of scope. The interested reader can consult [44] to learn about CI/CD. Model monitoring will be discussed in the next chapter.

16.1   Sharing Code

The easiest, safest way to share code is with a tool like `GitHub`, which we have been using for this book. Properly documenting and maintaining a project's code repository eases the task of promoting the code to engineering. A `README` file should be present in the repo to provide an overview of the purpose and use of the code. This can include summaries of the important functions with their inputs and outputs. The `README` can also include testing procedures, which will be discussed in the next section. The code files should include thorough documentation to understand the purpose of the important variables and functions, and the overall purpose of the file.

The code will likely change after it is released to downstream users, including engineering and customers. This may be to capture additional features or bug fixes, for example. It is important to track each package of code for future reference. This practice is called *versioning* the software. Different code platforms, such as `GitHub`, provide functionality for managing the code release from a repository. The specific steps are outside the scope of this text, and they will likely change over time, but awareness of this process is important. Different customers might use, and even require, different versions of project code. It is essential to have a way to release different code versions, to track them, and to maintain them over time.

16.2   Testing

There are many kinds of testing in software development, and we will briefly review some pertinent tests for data science. *Unit testing* gives developers confidence that their code performs as expected. The process consists of writing *unit tests*, which are used to test discrete units of code such as functions. Writing tests for each small piece of code enables rapid testing, isolation of bugs, and the surfacing of unexpected results. For these reasons, it is good practice for data scientists to write unit tests when developing code. It is also important to include clear comments on what each test does. This will allow for verifying that the tests match expectations.

Testing the data and the model output is also highly recommended. This practice can help to identify issues that may arise when new data is collected, when the code evolves, or when the model is updated. Particularly in machine learning, it is possible for the model to receive malformed or intermittent data and return an incorrect result. In such an event, the bad result may go undetected since nothing breaks.

Testing the stability of model output is also essential. For example, a data scientist may train a model that performs well. She may proceed to refine the application code for release to engineering, while keeping the model unchanged. Confirming that the model produces the same predictions for test cases will be important. A set of test cases with model inputs and outputs (e.g., predicted probabilities) can be created and saved with the code. Their careful design is essential, since their coverage defines the scope of testing. The tests may include:

- Common cases, such as frequent input combinations

- Cases where different input values are missing

- A case where all inputs are missing

- Edge cases which may be challenging for the model

- Cases for each predicted label value, when the task is classification

The practice of writing tests on the model output allows the data scientist to assert that the model predictions match known values, both at present and over time. Additionally, this allows others to test the application, such as quality assurance (QA) engineers. Since QA engineers are a step removed from the data science project, it will be harder for them to detect issues; the tests will provide a layer of support.

Python has modules for unit testing, and one of the modules is *unittest*. It provides a rich set of tools for constructing and running tests. This includes a set of methods for Boolean verification such as `assertEqual()`, which checks if two values are equal. For example, we might assert that a given model input will produce an expected output. The output may be known from an earlier call to the model. For the case where the prediction does not match the expected value, the test would fail and issue an `AssertionError`. The code block below illustrates two simple unit tests. The first test passes and the second test fails. An explanation follows the code.

```
from unittest import TestCase

# check if two statements are equal
actual1 = 100 % 2 == 0
expected = True

# create TestCase object
tc = TestCase()

# run test to assert equality, printing the result
# the test passes, returning None
print(tc.assertEqual(actual1, expected))

# run a second test, with different answer
# this will fail and throw an error
actual2 = 100 % 2 == 1
print(tc.assertEqual(actual2, expected))
```

The unittest module includes the TestCase class, which provides a number of methods to check for and report failures. In the first test, we compute an actual value which evaluates to True (since 100 divided by 2 has remainder 0). The expected value is also True. The assertEqual() method checks that the actual value matches the expected value. Since these values match, the test passes. For the second test, we compute an actual value which evaluates to False. The assertion checks if False is equal to True, which is a false statement. The output includes the error:

```
raise self.failureException(msg)
AssertionError: False != True
```

Upon reviewing the output, the data scientist can investigate the code, correct the error, and rerun the code. This can be repeated as necessary until all of the tests pass.

The unit tests above used simple arithmetic to produce the *actual* values. In practice, a more common way to generate actual values will be to pass the inputs to the model and run inference (make the predictions). Next, the relevant output can be used for the actual values. For a classification task, the actuals might be predicted probabilities. For a regression task, the actuals will be the predicted target values.

There are times when a test fails because it is written improperly. The developer needs to carefully review the tests and run them, to ensure they meet expectations. This may sound circular or excessive, but it can happen and faulty logic is often the culprit. One way to mitigate these errors is to design tests consistently. Specifically, testing some functions with assertEqual() and others with assertNotEqual() can lead to confusion. It can be better to design the latter tests to check for equality.

16.3 Containers

After the data science team completes a model, it will be delivered to engineering. It will be staged in a testing environment to ensure that everything works properly. From there, the model will be promoted to a production environment for deployment. Oftentimes when running the code on another machine, there may be compatibility issues which cause failure. There may be a difference in Python versions or Python modules, for example. It can take great effort to keep all machines running the same package versions (consider some organizations use dozens or more machines). Moreover, it may not even be possible to sync all versions. For example, two different ML applications may require different versions of a module.

The *container* emerged as a tool to maintain consistent, isolated project environments. It is a standalone, executable package containing everything needed to run an application. This will include software, configurations, and variables. The package can then easily be shipped to another machine, and run on that machine. It is even possible for a single machine to run multiple containers with different package versions. For example, a cloud server might run several machine learning models at the same time.

Virtualization is technology that uses software to simulate hardware functionality, thereby creating a virtual computer system. Containers are designed to share the operating system of a machine, rather than requiring their own copies. This is particularly helpful when multiple containers run on the same machine. Based on this feature, we say that containers are *lightweight*.

The universally accepted, open-source tool for containerization is *Docker*. Getting started with Docker includes learning some of the terminology and basic commands, which we outline next. A *Docker image* is a read-only file that contains instructions for creating a container. We can think of the image as a template. New images can be created, and they can also be shared. The online repository *DockerHub* has a large array of Docker images for the community (see `dockerhub.com`). It is similar to `GitHub` for Docker images.

A *Docker container* is a runtime instance of a Docker image. As we will see shortly, it is created by running the `docker run` command. A Docker container can run on any machine that has Docker installed. This allows great portability, as Docker can run on machines including a laptop, local server, or cloud server. Docker runs on operating systems including Linux, Windows, and Mac.

16.3.1 Installing Docker

For current Docker installation instructions and other details, please see the official Docker documentation here: `https://www.docker.com/`. Many organizations will opt for installing Docker on a Linux machine running in the cloud. This pathway will require running steps at the command line. Details can be found on the Linux installation page. For the purpose of this demonstration, we will install and run Docker Desktop for Windows. The steps involve downloading an executable file and following the configuration steps in the wizard. The steps for installing on a Mac are very similar.

After the Docker installation completes, we can test that it installed by opening a terminal and typing `docker` at the command line. Figure 16.2 shows some of the output in PowerShell. It shows that Docker is found, and there is a listing of common commands. If Docker is not found, the install can be rerun.

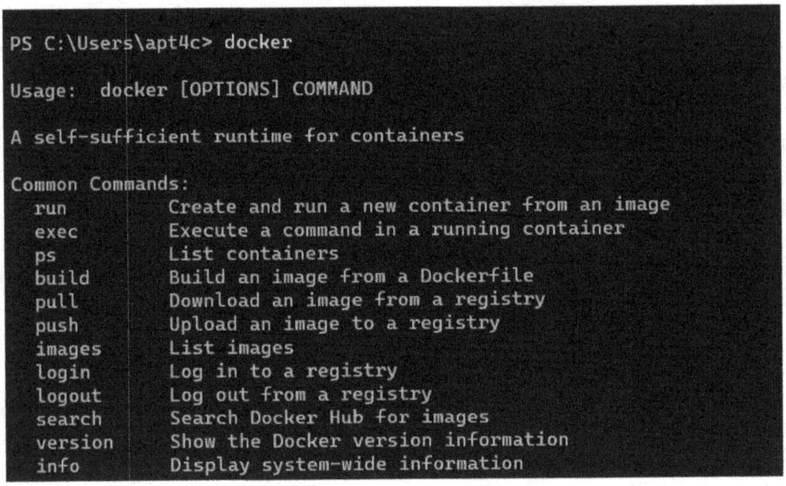

```
PS C:\Users\apt4c> docker

Usage:  docker [OPTIONS] COMMAND

A self-sufficient runtime for containers

Common Commands:
  run         Create and run a new container from an image
  exec        Execute a command in a running container
  ps          List containers
  build       Build an image from a Dockerfile
  pull        Download an image from a registry
  push        Upload an image to a registry
  images      List images
  login       Log in to a registry
  logout      Log out from a registry
  search      Search Docker Hub for images
  version     Show the Docker version information
  info        Display system-wide information
```

FIGURE 16.1: Checking for Docker

16.3.2   Common Docker Commands

Working with Docker requires running various commands in a CLI. A brief description of some of the commands follows. We will practice the commands later when we step through the process of building and running a Docker container.

command	purpose
docker build	build an image from a Dockerfile
docker run	run a container from an image
docker images	list all images downloaded
docker ps	list all running containers
docker stop	stop a running container
docker rm	remove a stopped container

16.3.3   Dockerizing an ML Application

In this section, we will walk through the steps of running a machine learning model from a Docker container. As a prerequisite, we need a Python script for the application. Then we can Dockerize the application by following these steps:

1. Create a `requirements.txt` file
2. Create a Dockerfile
3. Build the Docker Image

The `requirements.txt` file indicates the modules required to run the application, with their versions included. This removes ambiguity from the dependencies. In fact, even if Docker is not used, the `requirements` file should be included with the project.

A Python script named `housing_script.py` will be used for this illustration. The code was taken from the linear regression modeling chapter from earlier, and it is shown below. The steps include splitting the data into training and test sets, fitting a linear regression model, and printing the R-squared measured on the test set.

```python
# load the modules
from sklearn.datasets import fetch_california_housing
from sklearn.model_selection import train_test_split
from sklearn.preprocessing import StandardScaler
from sklearn.linear_model import LinearRegression
from sklearn.metrics import r2_score

import pandas as pd

housing = fetch_california_housing()

x_train, x_test, y_train, y_test = train_test_split(
                        housing.data,
                        housing.target,
                        train_size=0.6,
                        random_state=314)

reg = LinearRegression().fit(x_train, y_train)

y_test_pred = reg.predict(x_test)

r2 = r2_score(y_test, y_test_pred)
print('R-squared:', r2)
```

We can test the script by opening a terminal, navigating to the directory with the script, and running the command:

```
python housing_script.py
```

The output contains the R-squared as expected:

```
R-squared: 0.6092116009090456
```

For this simple model, the only module in use is scikit-learn. We include it in the requirements.txt file with its corresponding version:

```
scikit-learn==1.0.2
```

Next, we create the Dockerfile to set up the Python environment for running the script. Note that the file name does not have an extension. The Dockerfile uses specific commands, which are explained below.

```
FROM python:3.9
WORKDIR /src
COPY requirements.txt .
RUN pip install --no-cache-dir -r requirements.txt
COPY . .
CMD ["python","housing_script.py"]
```

Line 1 specifies that Python 3.9 is the base image.
Line 2 sets the working directory to /src.
Line 3 copies the requirements.txt file to the working directory.
Line 4 installs the Python packages in requirements.txt using pip.
Line 5 copies the files in the current directory to the working directory.
Line 6 runs the file housing_script.py using Python.

Lastly, we will build the Docker image. We can first verify that Docker is running by opening a terminal and running this command:

```
docker info
```

If Docker is not running, an example error like this might appear in the output:

```
Server:
ERROR: error during connect: this error may indicate that
the docker daemon is not running...
```

Once we verify that Docker is running, we can navigate to the directory with the Dockerfile and run this command:

```
docker build -t housing .
```

where **housing** is the name given to the image. Note the dot . at the end of the command to build in the current directory. As the image builds, logging will stream to the console. Here is an example of a successful build message:

```
[+] Building 85.4s (10/10) FINISHED
```

We can list the Docker images with the command

```
docker images
```

The output shows the **housing** image with its unique identifier (IMAGE ID) and size:

REPOSITORY	TAG	IMAGE ID	CREATED	SIZE
housing	latest	e3038e40147f	44 minutes ago	1.3GB

We can build and run a new container from the image using the command

```
docker run housing
```

The output is the same R-squared that we encountered when we ran the Python script earlier:

```
R-squared: 0.6092116009090456
```

This validates that Docker has created a container and run the machine learning script successfully in that container. We can see a list of all containers with the command

```
docker ps -a
```

Figure 16.2 shows the results following the creation of three containers based on the **housing** image. The output includes a unique identifier (CONTAINER ID) and a randomly created name for each container, among other information.

```
CONTAINER ID   IMAGE     COMMAND                 CREATED            NAMES
2e7bec864701   housing   "python housing_scri…"  About a minute ago   practical_blackburn
4fa0261c44e4   housing   "python housing_scri…"  3 minutes ago        frosty_jepsen
e335cd66b2a7   housing   "python housing scri…"  20 minutes ago       intelligent lederberg
```

FIGURE 16.2: Listing of Docker containers

In practice, after the image is created, it can be sent elsewhere to run on a different machine. An engineer might test the model on a cloud server, for example. She would download the image, create the container, and run it using the commands that we reviewed.

As the number of ML applications grows, the number of supporting containers may grow along with it. Managing and orchestrating containers becomes an important requirement and it can present a challenge. *Docker Compose* is a tool for defining and sharing multi-container applications. It allows for managing a system of Docker images in a single file. Additionally, there are cloud tools for container scheduling and orchestration, such as Amazon Elastic Container Service (ECS).

16.4   Chapter Summary

Deploying models and making them reproducible can be challenging. There are many steps required, but there are several practices and tools that can help. This chapter outlined the testing of code, data, and output. It also provided an introduction to containers.

The easiest, safest way to share code is with a tool like GitHub. Carefully documenting and maintaining a project's code repository eases the task of promoting the code to engineering. Over time, the code will change, and different users may have different versions. It is important to track each package of code for future reference.

Writing a comprehensive set of tests takes time and care. Incorrect tests and missing tests will compromise the entire exercise. The tests should be reviewed for proper coverage and execution. They should include common cases and edge cases, among others.

Providing a container with a data science project allows others to more easily replicate the work. Containers can hold all of the dependencies and configurations for an application. They can be used to stand up, run, and tear down a temporary, isolated environment. This can accelerate the path to production. Docker is currently the most popular container software. For more Docker details and examples, one helpful source is [45].

Now that we have covered some of the important steps in model deployment and reproducibility, we will review model risk management and model monitoring. Model risk is something that should be considered throughout the model lifecycle, starting with model conception.

16.5   Exercises

Exercises with solutions are marked with Ⓢ. Solutions can be found in the book_materials folder of the course repo at this location:

https://github.com/PredictioNN/intro_data_science_course/

1. List three things that must be done to put a model in production.

2. Ⓢ Explain why it is useful to provide unit tests with data science applications.

3. Ⓢ Investigate *sanity tests* and explain their value.

4. Investigate *regression tests* and explain their value.

5. Ⓢ Explain the difference between a Docker image and a Docker container.

6. A data scientist has developed an ML application in Python and wants to allow others to easily run it. Provide an outline of the steps that should be taken.

7. Ⓢ Is it possible to provide a name for a container? If so, how can this be done?

8. Ⓢ A data scientist runs a container that rapidly makes predictions using a model and then stops running. She checks the list of containers by running the command `docker ps` but does not see the container listed. Does this make sense? Provide justification for your answer.

9. Modify the `housing_script.py` file and run it in a Docker container. Feel free to use a completely different model, but be sure to print a result.

10. When running Docker, the logging mentioned a *daemon*. Investigate what a *daemon* is and provide a brief summary.

17

Model Risk

We dedicated a good amount of study to collecting data, setting up models, fitting the parameters to data, and interpreting results. What can go wrong? A lot, it turns out. Let's begin with a definition of model risk from the Board of Governors of the Federal Reserve System's Guidance on Model Risk Management:

The use of models invariably presents model risk, which is the potential for adverse consequences from decisions based on incorrect or misused model outputs and reports.

For the full report, visit:

```
https://www.federalreserve.gov/
supervisionreg/srletters/sr1107.htm
```

We can think about the entire process as the model. In this way, we consider not only the risk from, say, a mathematical or statistical model, but also the risk from the systems running the model and the input data. Each of these components can also introduce risk that needs to be managed.

Some divisions or entire companies make decisions based on models running in production. Incorrect predictions can lead to poor decision making and financial loss. Biased data and models can cause reputational damage and systematic harm to individuals. Copyright infringement can lead to lawsuits from content owners. An enlightening book replete with hard-hitting case studies on model risk is *Weapons of Math Destruction* [46].

Some companies never recover from bad models. One doesn't need to search for very long to find articles in the news about problematic models which led to financial meltdowns, disparate treatment, legal risk, and reputational damage. Here are some examples:

- Long-Term Capital Management was a hedge fund founded in 1994 which had a strong start and high-profile leadership. When markets moved against their quantitative strategy and massive leverage, the firm faced multibillion dollar losses and a collapse.

- Zillow used predictive models to fuel the purchase and flipping of homes in their Zillow Offers program. During the COVID pandemic, their predictions were highly inaccurate, leading to losses in excess of $400 million and a 25% reduction in staff.

- Stability AI developed a model to create images from a text description. The firm is currently embroiled in a lawsuit with Getty Images, which claims that Stability AI unlawfully scraped millions of images from its site. Getty Images believes that Stability AI used these images in training its model.

The purpose of this chapter is not to scare the reader away from data science, of course, but to bring awareness of what can go wrong and how to prevent it. I spent the early part of my career building models for investment banks and asset managers as a Quant. One of my roles a bit later was in model validation, where I examined, tested, and challenged an array of models for a major commercial bank. My group was independent from model development, mandated by the federal government, and designed to provide another effective line of defense against model risk. The work that my team did gave me a tremendous appreciation for managing model risk, and it made me a better model developer. It opened my eyes to the "gotchas" hidden in models, such as predictors that might lead to unfair treatment, assumptions that might not hold up in practice, and patterns that likely would vanish.

Model validation is an essential function that is not restricted to banks or the financial sector. Anywhere that a model is used, it introduces risk that needs to be measured and managed. The person doing this work must review the model carefully to ensure proper use, correct design and logic, proper data handling, robust performance measurement and testing, and complete documentation. From the standpoint of a model developer, while validation introduces extra work, it brings the benefit of a fresh perspective. An effective validation will challenge the model and likely strengthen it.

17.1 Model Documentation

Model documentation is essential for communicating information about the development process, the analysis conducted, and the decisions made. The model developer should create a *model development document* or *devdoc* to capture the important elements in the process including:

- Model purpose

- Names and titles of stakeholder

- Data assumptions and limitations

- Model assumptions and limitations

- Data description: time period, context on the variables

- Description of the model

- Process diagram

- Data processing, including steps to create predictors

- Outline of data collection

- Justification for predictors included in model

- Justification for predictors excluded from model

- Model outcomes analysis, such as backtesting

- Model sensitivity testing

- Model benchmarking

- Monitoring plan

- Summary of failed experiments

A model validator will find the devdoc extremely useful in gaining a better understanding of the end-to-end process. She can then write her own report on findings from the validation process. For the lifetime of the model, the documentation will grow and evolve to reflect the latest state.

It is helpful to summarize the high-level details of a model in one place, and the *model card* is such a tool (though this name is not universal). Historically, organizations that developed models were typically the only ones to use them. This is no longer the case. Increasingly, models are a shared resource. A principal reason for this is cost: building state-of-the-art models, such as models that understand language, requires massive datasets and expensive computing resources. It also requires specialized expertise. Fortunately, many of these leading machine learning models are freely available on platforms such as Hugging Face (https://huggingface.co). Repurposing an appropriate model can provide large savings and accelerate development. An important question is whether a model is appropriate for a given use case, and this is where model cards can help. An informative model card can provide transparency and safety, which helps foster model use. In practice, the depth of information provided in model cards runs the gamut from empty to complete, but a card might include:

- Model description: the high-level purpose of the model

- Model background information

- Model uses and limitations

- Model architecture (e.g., logistic regression)

- Technical notes

- Instructions on how to use the model

- Model size (typically measured in number of parameters)

- Warnings of possible model bias and toxicity

- Information about the training data

- Performance on tasks using defined datasets, called *benchmark datasets*

- References to supporting papers

In addition to consuming models from a community platform, it is also possible for users to contribute their own models. Providing a detailed model card with the model is highly encouraged.

Whether a model originates inside or outside of an organization, model cards are a quick way to communicate the suitability, benefits, and risks of models. Creating them and keeping them current is essential to proper model risk management.

Next, we will step through the important areas of model validation. This will review some of the items listed above and explain why they are so important in mitigating and managing model risk.

17.2   Conceptual Soundness

When we consider a system to solve a problem with data, it is essential to validate each component: the data collection process, logic, models, sequence of steps, assumptions, and limitations. It is enlightening to create a detailed flow diagram to understand each step, possible gaps, bottlenecks, redundancy, and feedback loops (see Figure 17.1 for an example). The analyst should be asking questions such as:

- Why is this being done?

- Does this make sense?

- Is this the best way to solve the problem?

- Are there other alternatives that should be considered?

- Is this assumption/limitation reasonable? Could it cause problems later?

- What might cause this to break?

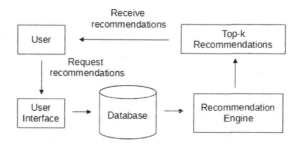

FIGURE 17.1: Sample Recommender Flow Diagram

The model is often where a good portion of the complexity and risk is located. This is because models are abstractions of reality where a tradeoff is made. Consider face-recognition algorithms. How can they still recognize someone with glasses or a slightly different hair style? The model does not attempt to match every pixel, but rather it learns a representation, or simplified structure, of the face. The model essentially condenses the information down into a subset of information which may be sufficient to solve the task at hand. This happens when converting an image in RAW (uncompressed) format to a JPEG, for example. Does it work? In some cases, it does. The JPEG is a lower-resolution image, but this is sufficient for sending photos on a mobile app; it won't be good enough for museum-quality prints, however.

The models used in data science and machine learning will make assumptions to similarly abstract away details to simplify the problem. We might assume that a variable follows a normal distribution, when this may not be exactly true. Does it matter if a variable isn't quite normally distributed? This depends. If the variable under study is the percentage change of a stock price, and the model will be used in a trading system, then it certainly matters. Suppose the real-world probability of a negative value is greater than what the normal distribution would suggest. This could lead to serious financial losses.

Each model brings a level of complexity which depends on factors like the architecture and the number of predictors. We should strive to use only the level of complexity necessary to solve the problem. There may be a temptation to use the most advanced methods to solve a problem, but if it can be solved with linear regression, by all means use this method. However, if a non-linear relationship exists between predictors and a target, then a more flexible (and complex) model may be warranted. The data should be the guide, and not publication trends.

When I was a graduate student, one of my areas of specialization was regime-switching models. In the right circumstances, these models were very useful and powerful. They used additional parameters when compared to linear regression models. These parameters added flexibility, but also complexity.

When I proposed the model to practitioners, many expressed hesitation to use them in production. After several discussions, I found a common difficulty: for some kinds of data, the parameter estimates were very unstable. Over time, they would vary greatly, and the models produced inaccurate predictions. This led me to realize that much care and testing was required when introducing complex models.

This leads to the point that model developers need to justify their modeling decisions. This should be done through analysis and testing. More than one model should be considered, so that comparisons can be made. The final model does not need to be the optimal model over the universe of models, but it needs to make sense, provide demonstrable value, and carry an acceptable risk profile.

17.3   Data and Inputs

The data considered for the model, as well as the model inputs, are crucial to strong performance. First, there needs to be an understanding of what all of the data means. In the ideal case, a *data dictionary* will describe each of the variables to provide context. Unfortunately, a data dictionary may not be available. In this case, the data scientist might need to spend time searching for clues, documentation, and the right people to ask. For one consulting project, my team built a model without understanding all of the variables. We noticed that the model fit perfectly, which was a red flag that something was amiss. It turned out that one of the predictors was essentially the target variable. This kind of thing happens more than you might guess.

Suppose the data is now understood and documented. What else is important? If a variable will be an input or predictor in a model, we must ensure that:

- The variable will be readily available in the future. If the variable will need to be processed in real time, but the system requires two hours to process it, then we should not include the variable in the model.

- The variable does not induce bias

- The variable makes intuitive sense in the model

- The data is accurate. When possible, it should be cross-referenced.

- We are permitted to use the data. For example, unauthorized use of copyrighted material can provoke a lawsuit.

After reviewing the data and model inputs, it is also important to ensure that the dataset is sufficient. For example, if we are building a risk model, then the data should include observations from a period of heightened risk. For cyclical data, it is important to include at least one full cycle in the dataset.

17.4 Outcomes Analysis

At this stage, we have reviewed the model components, the data, and the predictors in the model. Next, we will want to understand the utility of the model outputs. The *outcomes analysis* step will compare model outputs to actual outcomes. We might investigate specific cases which arise frequently. Alternatively, we might make predictions on a batch of data and compute a metric, as we did when evaluating regression model performance.

The process of *backtesting* is a form of outcomes analysis that compares actual outcomes to model predictions from a sample time period not included in the training data. We might use historical data from the same period as the training data to evaluate performance. This is sometimes called *out-of-sample testing*. We might use recently collected data to measure model performance for *out-of-time testing*. This will give an idea of how the model will perform when it is deployed. Table 17.1 shows out-of-sample test results comprising predictions from model m_1 and actual outcomes.

TABLE 17.1: Out-of-Sample Test Results

index	prediction_m_1	actual
0	0.20	0.06
1	−0.10	−0.04
2	0.70	0.45
3	0.45	0.63
4	0.15	0.11

The metrics we have discussed earlier can be used to understand how the model is performing in aggregate. This may include adjusted R-squared and RMSE for assessing fit for linear regression models. For logistic regression models, precision, recall, and F1 score may be useful. Metrics should also be collected over time for *ongoing model performance monitoring*, which we will discuss shortly. Based on the out-of-sample errors in this case, the RMSE is 0.15. This can be verified by computing the squared errors between each prediction and its associated actual value, computing the average squared error, and taking the square root.

17.5   Model Benchmarking

The term *benchmarking* in model validation refers to comparing the model under consideration to an alternative model or metric. The benchmark model is generally a model with minimal complexity, which could have been developed with less effort. In fact, the supervised models that we have studied, linear regression and logistic regression, are good benchmark models. If one were validating a more complex model like a neural network, it would make sense to see how its performance compares to the appropriate regression model.

Sometimes very simple heuristics are available for benchmarking. A common question is to ask if a guess "is better than a coin flip." That is, there are cases where there is a 50% chance of simply guessing correctly (say on a True/False question). A useful model in this case would need to have accuracy better than 50%. One of the most famous prognosticators of all time, Punxsutawney Phil, is a groundhog who uses his shadow to predict if winter will get an extension. From the National Weather Service [47], it appears he has been correct about 40% of the time. This suggests his accuracy is worse than a coin flip, and we would do better by flipping his prediction.

Let's think about some benefits that a benchmark model can provide. First, it is possible that the performance of a benchmark model is similar or better than a more complex model. In this case, the benchmark model might be promoted as the *champion model* to be used in production. Second, the benchmark model can be used as a sanity check and to validate different inputs. Suppose the benchmark model and the model under consideration are given the same input and they produce vastly different output. This provides the opportunity for research. The analyst might be able to explain this difference based on the different frameworks, or she might find a bug in the code. In either case, diving deeper should provide a better understanding.

Table 17.2 repeats the m_1 out-of-sample test results alongside predictions from a benchmark model m_{bm}. The benchmark model is a simple linear regression model, while m_1 has greater complexity. For each observation, the prediction from m_{bm} is closer to the actual value. Additionally, the RMSE is smaller for m_{bm} (0.03) than m_1 (0.15). This provides evidence that m_{bm} is the more accurate model. If the pattern holds for a larger sample of observations, this would indicate that the more complex model is not warranted.

TABLE 17.2: Model Benchmarking: Out-of-Sample Test Results

index	prediction_m_1	prediction_m_{bm}	actual
0	0.20	0.05	0.06
1	−0.10	−0.03	−0.04
2	0.70	0.50	0.45
3	0.45	0.61	0.63
4	0.15	0.09	0.11

17.6 Sensitivity Analysis

The U.S. News and World Report is one of the most followed college ranking systems in the United States. Between 2021 and 2022, Columbia University plummeted from a ranking of 2 to a ranking of 18. This was caused largely by inaccurate data, and it prompted questions about the validity of a system with such a large sensitivity to inputs. Was the model faulty? This is difficult to say without an audit of the model and data, and I'm not here to criticize the approach. I do want to point out the importance in assuring that small changes to modeling assumptions do not lead to large changes in outputs. If a small change to a parameter or the addition of a predictor will result in massive changes to outputs, this can lead to backlash.

An important practice in model validation is conducting *sensitivity analysis*, which quantifies how a change in modeling assumptions translates to a change in output. In turn, changes to output will likely impact performance metrics. There are two pieces here that I wish to expand upon: the modeling assumptions, and the measurement of output. Let's discuss them in turn.

When we think about modeling assumptions, this may include things like assumed system dynamics, distributions of random variables, relationships between variables, parameters, and hyperparameters. The thoughtful developer considers each of these elements, makes decisions, and builds the model to express the decisions. The developer may have decided that a certain hyperparameter should have a value of 10, and perhaps tested values both near and far from 10. The developer might have considered different distributions for predictor X, deciding that a normal distribution is more appropriate than a t-distribution. Let's list some of the modeling choices we have encountered so far:

- When we split the data into a training set and a test set, we used a seed for the random selection of rows. A different seed likely would have split the data differently.

- We had a choice of methods for imputing missing data. For example, we might have imputed with the median or mean of a variable.

- When we used a linear regression model, we assumed that the errors were normally distributed

- When we used a linear regression model, this assumed a linear relationship between the predictors and target variable

To illustrate a real-world modeling assumption, I will draw from my first job as a credit derivatives researcher. To study different structures, we built financial models that took inputs such as the probability of default for various types of bonds. From careful testing, we learned that a small change to the default assumptions could produce a large change in predictions. In fact, a small change could be the difference between a lucrative investment and financial distress.

In measuring the impact on the output, we will need to decide exactly what should be measured. This might mean running the model on some important cases, such as interest rate forecasts for predetermined scenarios. It might mean running the model on the entire test set. The decision will depend on what is important to stakeholders and the business case.

Once the analyst has identified what should be changed (e.g., a seed, a variable's distribution, an assumed probability value), and what output should be measured, she can design the tests. Running the tests will produce output that might be supplemented with metrics, visualizations, and alerts. For example, an alert might be triggered if any test causes a change of 10% or more for any metric. The results can be reported to the appropriate stakeholders.

Table 17.3 shows sensitivity testing on m_1 given some model changes. The *base case* column shows predictions from the original model. The *chg seed* and *median impute* columns show predictions when a different seed and a different impute method were used, respectively. Note that the changes to predicted output were very small, which provides evidence of model stability. Accordingly, the changes to the RMSE statistics will also be small.

TABLE 17.3: Sensitivity Testing: Out-of-Sample Predictions

index	base case	chg seed	median impute
0	0.20	0.21	0.20
1	−0.10	−0.10	−0.11
2	0.70	0.70	0.71
3	0.45	0.44	0.45
4	0.15	0.14	0.14

17.7 Stress Testing

For certain kinds of models, it will be appropriate or even necessary to run them on plausible, extreme scenarios to measure the output. The scenarios often represent adverse conditions, although it is not required. We might predict a stock's performance in strongly and weakly performing markets, for example. The difference between *stress testing* and sensitivity analysis is that the former makes material changes to inputs, while the latter makes small changes to modeling assumptions. We expect the outputs to change dramatically when running stress tests, and we use the information to prepare and consider if the model is functioning properly.

For an example of stress testing, we might apply a financial model to estimate the impact of a dot-com crash or an inflation surge on an investment portfolio. To be sure, a future event won't mimic a historical event, but it can give a sense of the magnitude of risk. After building an inventory of scenarios and applying the model, there will be a set of associated outcome estimates as in Table 17.4. The analyst should question whether the results make sense. For example, if a historical event resulted in a 37% loss, and the model is predicting a 10% gain, this should elicit suspicion that the model may be problematic. Perhaps the model is not specified correctly for extreme inputs, and its structure should be reconsidered.

When the analyst is satisfied that the model is functioning properly for the stress tests, the calculation, testing, and reporting processes can be automated. Regular review of the outcomes should be conducted as part of ongoing monitoring (to be discussed next). Additionally, the stress test scenarios will likely need to be updated during the model lifecycle.

TABLE 17.4: Sample Stress Tests

scenario_id	scenario_name	description	outcome
1	Market Up	S&P 500 up 20%	+25%
2	Market Down	S&P 500 down 20%	−25%
3	Volatility Up	CBOE VIX up 20%	−15%
4	Volatility Down	CBOE VIX down 20%	+15%
5	Dot-com Crash	Repeat of dot-com bubble	−32%

Now that we have discussed both sensitivity analysis and stress testing, we can revisit the Columbia University case. If one or more of the model inputs between 2021 and 2022 changed drastically, this might have been more like a stress test than sensitivity analysis. This could explain the large change in ranking (but don't quote me on this).

17.8   Ongoing Model Performance Monitoring

Models need to be monitored to ensure that operational and predictive performance meet expectations. *Operational performance* includes response time and uptime, for example. As a reminder, the response time is the elapsed time from when the request is made to when the client receives the first byte of the response. The uptime is the percentage of time that the system is available and working properly. Predictive performance will quantify prediction errors using metrics such as recall and precision. Recall will measure the fraction of positive labels that were detected by a classifier, while precision will measure the fraction of predicted positives which were in fact positive. A data scientist will generally focus on predictive performance.

Before a model is placed into production, a *performance monitoring plan* should be drafted and agreed upon by the key stakeholders. There should be a governing body which provides oversight and has the authority to enforce the plan, should action be needed. The plan should include but is not limited to these elements:

- The metrics to be monitored

- For each metric, the thresholds that should trigger an alert

- The next steps when a threshold is crossed. This should mention stakeholder responsibilities.

- The review frequency

- The name, title, and voting authority of each person on the model review committee

Here is a small, fictitious example which elaborates on the elements listed above:

START OF PERFORMANCE MONITORING PLAN

Model ID	123
Model Name	Student Loan Default Model
Date of Report	2023-04-15
Period Covered	2023-01-01 through 2023-03-31
Review frequency	quarterly

METRICS, STATUS, AND THRESHOLDS

Metric	Alert Level	Threshold
F1 score	YELLOW	< 0.8 or drop exceeding 10%
F1 score	RED	< 0.7 or drop exceeding 20%
. . .	. . .	. . .

ACTIONS AND ESCALATION POLICY

In the event that the YELLOW threshold is crossed, these stakeholders will be alerted:

Product Manager
Engineering Manager
Data Science Manager

The model development team and the engineering team will review the model components and data. The model may be retrained to include recent data.

In the event that the RED threshold is crossed, these stakeholders will be alerted:

VP of Product
VP of Engineering
VP of Data Science
Product Manager
Engineering Manager
Data Science Manager

The model will be taken out of production so that it can be redeveloped. The model development team and the engineering team will review the model components and data.

ROLES AND RESPONSIBILITIES

Role	Name	Title
Model developer	John Smith	Data Scientist
Manager of model development	Jane Cooper	Data Science Manager
Model validator	Xie Xu	VP, Model Validation
Manager of model validation	Carlos Juarez	SVP, Model Validation
. . .	. . .	. . .

MODEL GOVERNANCE COMMITTEE

Name	Title	Department	Voting
Dan Johnston	Manager	Student Loans	Yes
Cindy Oh	SVP, Model Governance	Model Validation	Yes
Monica Swift	Head of Risk	Risk Management	Yes
...	...	...	...

END OF PERFORMANCE MONITORING PLAN

To enable effective monitoring, the metrics need to be calculated and reported in a timely fashion. Ideally, there is a reliable, automated system in place. When the model is initially launched, ideal thresholds may not be known. This may take experimentation, as an overly stringent threshold may trigger false alerts. The monitoring plan and system should be flexible to allow for adjustment.

Finally, it is important to interview the model users to understand if things are performing as intended. For example, if a model seems to make a common error, the model users might override the model with human judgment. If this information is captured, it may be possible to improve the model and eliminate the override.

17.8.1   Diving Deeper on Monitoring

We learned that metrics will be monitored over time to understand the quality of predictive performance. What happens when metrics take a tumble and trigger a review? It will be important for analysts to dive deeper, and there are several things that can be reviewed.

Two important things that should be implemented for monitoring are making the system *observable* and providing relevant *baselines*. An observable system will capture, or log, the important steps and intermediate values for later review. The logging can reflect samples of model inputs and outputs, for example. It might reflect steps where code branching occurs, so that analysts can understand if the code took an unexpected turn. Everything is captured in real time. This level of observability will make it easier for data scientists and data engineers to spot problems, as it means less post hoc hunting and testing.

To monitor model performance over time, it is useful to first establish baselines. The baselines might include the F1 score measured on the test set, for example. Additionally, important statistics of the predictors might be included in the baseline, such as five-number summaries. Here is a small example of a JSON object which stores baseline statistics for a given date. The baseline data comprises one metric (F1 score) and a five-number summary for the variable age.

```
{
    'baseline_date': {
       '2023-01-31':
          {
            'metrics': {
                      'F1 score': 0.68
                      },
               'predictors': {
                      'age':
                             {'min': 25,
                              'q1': 30,
                              'median': 37,
                              'q3': 55,
                              'max': 95}
          }
      }
   }
}
```

The baseline information may be stored in a data lake or managed with a cloud-provider service such as Amazon SageMaker. It can be accessed and used programmatically.

In the event that metrics trigger an alert, data scientists can start leveraging the logs and the baselines. They can first check the logs for indicators of problems. Next, they can investigate if there was a change in the distribution of the predictors, for example. To statistically test if the distribution of a variable changed between times t_1 and t_2, the two-sample Kolmogorov-Smirnov test (K-S test) could be used. The K-S test is beyond the scope of this book, but more details can be found in [30].

A change in a predictor's distribution over time is called *factor drift*, and it can be problematic. This is because the model learned specific relationships between predictors and a target from a training set. If the distribution of a predictor changes, its relationship with the target may change as well. We would need to retrain the model with the new data to capture the change. There is no guarantee, however, that the predictor will continue to provide the same level of predictive ability.

Other potential areas for investigation occur earlier in the data pipeline. There may be defects in the processed data, the raw data, or the data feeds. Perhaps there was a step where an integer was expected, but a string was input. If the code was not properly written to handle such a case, an issue may arise. There might have been a period of time when a certain data feed wasn't working properly, and an important predictor was not supplied to the model. Diving deeper into the data can take a lot of time and effort, and this reinforces the importance of system observability.

17.9   Case Study: Fair Lending Risk

This section will provide a detailed, computational example of *fair lending risk*. We will start with some background on the topic.

17.9.1   Fair Lending Background

Lenders such as commercial banks offer loan products to customers. For many, this is a major pathway for purchasing a car or home, or affording college. The lender will decide which applicants should be approved for a loan (the *underwriting decision*) and what interest rate should be offered (the *pricing decision*). Lenders want to avoid the possibility of nonpayment, or default, on part of the loan or the full loan. They use historical data and judgment, among other things, to make lending decisions. Decisions which lead to rejecting applicants or offering a higher interest rate on the basis of things like race will promote unfair lending practices.

Unfortunately, some lenders include bias – knowingly and sometimes unknowingly – in their decision. There are many ways that this can happen. In one pathway, human decisions are biased against certain groups of individuals. If this data is used to train a model, the model will learn the patterns and systematically propagate unfair lending practices. A second possible origin is data imbalance. For example, there may be lopsided quantities of male and female loan applicants at a bank. There are tools for identifying and remediating bias and unfairness in data and models, such as SageMaker Clarify from AWS [48].

Federal bank regulatory agencies such as the Office of the Comptroller of the Currency have developed approaches to address unfair and deceptive lending practices. The Fair Housing Act and the Equal Credit Opportunity Act protect consumers by prohibiting unfair and discriminatory practices. Discrimination in credit and real estate transactions is prohibited based on factors including:

- Race or color

- National origin

- Religion

- Sex

- Age (provided the applicant is old enough to enter into a contract)

A *protected class* refers to groups of individuals protected by anti-discrimination laws, such as:

- Women

- Individuals over the age of 65

- Black, Hispanic, and Native American individuals

Illegal, *disparate treatment* occurs when a lender bases its decision on one or more discriminatory factors covered by fair lending laws. For example, a bank using a lending process where females are offered an auto loan with a higher interest rate than males would be a violation of fair lending laws.

In modern finance, banks and other lenders typically use machine learning models to make lending decisions. The underwriting decision may use a logistic regression model. Given approval, the pricing decision may use a linear regression model.

The model predictors should capture the ability and willingness of borrowers to repay the loan. The predictors should not use protected class information, as this can discriminate and promote unfair lending.

There will be two models discussed in our example. The underwriting model M_U will be used for making automated underwriting decisions. The fair lending model M_{FL} will be used to assess if the underwriting process was systematically fair. Given this background information, we will now review the data and modeling.

17.9.2 Numerical Example

We will explore fair lending risk with a synthetic mortgage lending dataset. The data can be found here:

`semester2/datasets/mortgage_lending.csv`

The fair lending notebook can be found here:

`semester2/week_15_16_fair_lending_application/`
`IDS_hw4_fair_lending_intro.ipynb`

This dataset includes a subset of what a typical file would look like. It includes:

- The demographic variable: `gender`

- The credit factor `fico` (for FICO score)

- The credit factor `loan_to_value`

- The loan attribute `loan_term`

- The loan attribute `rate_type`

- The target `denied`, where 1=denied, 0=approved

A higher FICO score reflects a lower risk of default. Loan-to-value ratio is the amount of the loan divided by the value of the asset. A higher ratio is considered riskier, as this implies a smaller downpayment.

Let's begin by importing modules and looking at some of the data:

```python
import numpy as np
import pandas as pd
import statsmodels.api as sm

syn = pd.read_csv('../datasets/mortgage_lending.csv')
syn.head()
```

OUTPUT:

id	fico	ltv	gender	occupation	denied	loan_term	rate_type
0	696	62	m	salesman	0	30	fixed
1	752	61	m	salesman	0	30	fixed

Note that `loan_to_value` is abbreviated as `ltv` due to margin constraints.

We conduct some exploratory data analysis to better understand the data and uncover potential predictors. It will be helpful to understand the composition of the records by occupation and gender, for example. We can build a two-way table to explore this. For example, of the 12 accountants in the dataset, 5 are female and 7 are male.

```python
pd.crosstab(syn.occupation, syn.gender)
```

OUTPUT:

gender	f	m
occupation		
accountant	5	7
contractor	1	4
...	...	...
lawyer	7	5
librarian	0	1
mason	0	1
mechanic	0	2
nurse	10	0
...	...	...

An important observation is that some occupations are comprised of a single gender. For example, there are 10 nurses, and all of them are female. Suppose nurses historically defaulted on loans at a high rate in the bank's dataset. If `occupation` is used as a predictor in a lending model, the model will learn that nurses have a high risk of default. This will lead to the model denying loan requests to nurses in the future, which effectively denies lending to more females. The `occupation` variable is acting as a discriminatory variable in this case, and it should not be included in the model.

Let's examine distributions of `fico` and `loan_to_value` for denied and approved applicants. These are summarized as boxplots in Figures 17.2 and 17.3. Denied applicants tended to have lower FICO scores and higher loan-to-value ratios, which makes sense.

```
sns.boxplot(x='denied', y='fico', data=syn)
sns.boxplot(x='denied', y='loan_to_value', data=syn)
```

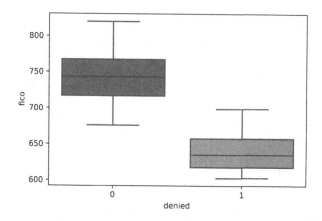

FIGURE 17.2: Denied by FICO Score

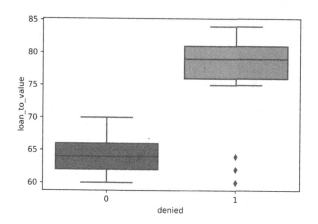

FIGURE 17.3: Denied by Loan to Value

Suppose the predictors used in the underwriting model M_U were

```
L_U = ['fico','loan_to_value','loan_term','nurse']
```

We will use these predictors to build the model next, but first: which of these predictors should have been used? The variables fico and loan_to_value measure credit worthiness and the ability to repay the loan, so these should be used. The variable loan_term is a loan attribute, and it should be used. As discussed earlier, nurse will introduce the risk of discriminating against females, and it should NOT be used.

Now let's train the underwriting model on the synthetic data.

```
# build nurse indicator variable
syn['nurse'] = syn.occupation.apply(lambda x: 1 if x == 'nurse'
                                            else 0)

X = syn[['fico','loan_to_value','loan_term','nurse']]

# include intercept term
X = sm.add_constant(X)

# target variable
y = syn['denied']

# train logistic regression model
result_uw = sm.Logit(y, X).fit()

print(result_uw.summary())
```

TABLE 17.5: Underwriting Model Parameter Estimates

	coef	std err	z	$P > \|z\|$	[0.025	0.975]
const	74.9067	4.66e+04	0.002	0.999	−9.12e+04	9.14e+04
fico	−0.1400	0.057	−2.455	0.014	−0.252	−0.028
loan_to_value	0.0795	0.113	0.706	0.480	−0.141	0.300
loan_term	0.4942	1552.666	0.000	1.000	−3042.675	3043.664
nurse	2.7672	1.892	1.462	0.144	−0.942	6.476

The parameter estimates are shown in Table 17.5. The only significant predictor in the model is fico as indicated by its p-value less than 0.05. The negative coefficient estimate of −0.14 indicates that a higher FICO score is associated with lower odds of denial, which makes sense.

Next, we train the fair lending model M_{FL} to understand if **gender** is a significant predictor. The model will include an intercept term, the credit factors, the attributes of the loan, and **gender**. If **gender** is a significant predictor in M_{FL}, this would indicate that it is being used to make lending decisions, which is discriminatory practice.

```
# convert categorical gender to indicator
syn['gender_female'] = syn.gender.apply(lambda x: 1 if x == 'f'
                                                  else 0)

X = syn[['fico','loan_to_value','loan_term','gender_female']]
X = sm.add_constant(X)

y = syn['denied']

result_fl = sm.Logit(y, X).fit()

print(result_fl.summary())
```

TABLE 17.6: Fair Lending Model Parameter Estimates

	coef	std err	z	$P > |z|$	[0.025	0.975]
const	96.2464	4.21e+04	0.002	0.998	−8.25e+04	8.27e+04
fico	−0.1670	0.006	−2.516	0.012	−0.297	−0.037
ltv	0.1422	0.137	1.041	0.298	−0.126	0.410
loan_term	0.2092	1404.537	0.000	1.000	−2752.632	2753.050
gender_f	4.0350	2.200	1.834	0.067	−0.278	8.348

Note that **gender_female** is abbreviated as **gender_f** due to margin constraints.

The parameter estimates in Table 17.6 indicate that only **fico** is significant. The predictor **gender_female** has p-value 0.067. Since this value is close to 0.05, the process should be closely monitored over time for fair lending risk. In practice, all of the relevant protected class variables would be rigorously tested and examined.

17.10 Chapter Summary

By design, models simplify reality to capture salient aspects of a process. In doing so, they introduce model risk. The risk can cause financial, legal, reputational, and societal damage. We studied fair lending risk in detail, which results from disparate treatment of protected classes.

We reviewed how machine learning models are used in lending decisions, including underwriting and pricing. Diving into the data showed how discriminatory behavior can unintentionally creep into models. In this case, all of the nurses were female, and this could bias the lending practice. The case study illustrated the application of data science to a critically important task in finance. More broadly, it outlined the examination of one aspect of risk.

To properly validate a model, it must be effectively challenged. The conceptual soundness of the overall system and each component must be examined and understood. The validator needs to understand the assumptions and limitations of the data and model, as well as the development decisions made. A model development document should capture this information, and it should be updated over the model lifecycle.

The performance of the model needs to be measured and monitored over time to ensure that it meets expectations. Outcomes analysis will compare predicted values to actual outcomes. If model users find that predictions exhibit bias, or deviate from actual values in a predictable way, they might override the model. Model users should be interviewed to capture and understand any overrides, as this presents an opportunity to improve the model.

To make models easier to monitor, they should be observable, and their baseline statistics should be captured. The important steps and intermediate quantities can be logged. If metrics deteriorate, the logs can be searched first. Other layers to investigate are the predictors, processed data, raw data, and data feeds. The distributions of predictors can change over time, and this is called factor drift. This phenomenon can be problematic as the relationship between a predictor and target is fundamental to model performance.

Model benchmarking compares the model to an alternative model or metric. The benchmarks will be relatively simple approaches or heuristics. A more complex model should beat these benchmarks to be valuable. Sensitivity analysis is the exercise of slightly changing the model assumptions and measuring the impact to the output. Small changes are expected, and large changes will warrant further investigation. Stress testing is the practice of running the model on plausible, extreme scenarios to measure its response. This may surface issues with the model which can be investigated. It is also useful as a "what if" exercise.

We have covered a lot of ground in this introduction to data science. In the next chapter, we will plan some next steps for going deeper and broader. The field is extremely active – particularly in deep learning. We will see some resources for learning more, and some of the recent advances.

17.11   Exercises

Exercises with solutions are marked with Ⓢ. Solutions can be found in the book_materials folder of the course repo at this location:

https://github.com/PredictioNN/intro_data_science_course/

1. Ⓢ You are asked to validate a linear regression model. What might you verify as part of checking the conceptual soundness of the model?

2. You have discovered a powerful predictor, and its correlation with the target variable is 1. What is your next step?

3. Ⓢ Why is a benchmark model important?

4. You discover that the predictive performance of a logistic regression model is better than the champion neural network model. All else equal, what should you do?

5. Ⓢ A technology startup has developed a sophisticated model to predict fraud. The startup believes that the F1 score is the most important metric to its target customer, and through testing, their model has a very high F1 score. Assuming that F1 score is the only thing that matters to the customer, are they likely to use the model? Explain your answer.

6. Ⓢ Explain how stress testing is different from sensitivity analysis

7. A hedge fund is designing a set of stress tests. Select all of the tests that may be useful.

    a) The S&P falls by 1%
    b) The S&P falls by 25%
    c) An asteroid hits company headquarters
    d) Inflation skyrockets by 10%

8. You would like to conduct sensitivity analysis on a Lasso regression model. List some possible sensitivity tests that you could run.

9. Ⓢ Explain the difference between monitoring and observability. Hint: is there a dependency between them?

10. Why is it important to establish baselines for monitoring?

11. Ⓢ How might you detect factor drift? How could you mitigate factor drift?

12. Visit `http://huggingface.co` and review the model cards of some models. One good example is `bert-base-uncased`. Note anything that you find interesting.

13. Ⓢ A data scientist located a model on the company's production server that predicts credit card losses. Her team needed a credit card model, and so she started using it without reading its documentation. What is one of the largest risks of this action?

14. What is the practice of redlining?

15. Ⓢ Which of these variables are sensible when making lending decisions? Select all that apply.

   a) Salary
   b) Credit score
   c) Religion
   d) Loan purpose

16. Ⓢ A fair lending model indicated that gender was a significant predictor. What does this mean?

17. A quantitative analyst constructed a mortgage underwriting model to decide who should be approved for a mortgage. One of the variables which best explained mortgage defaults was zip code. Is this variable problematic from a fair lending perspective? Why or why not?

18

Next Steps

Congratulations, you've made it through the content of the book! You should now have a good understanding of the field of data science, important data literacy topics, and how to implement a data science pipeline. Data science, and machine learning in particular, is vast and growing rapidly. The intention of this chapter is to provide some ideas and suggestions of where to go next. Over time, we can expect to see new areas of activity, applications, and tools.

18.1 Building Blocks

To prepare for the more advanced models and applications, it will be necessary to add to your skills in mathematics, statistics, computing, data analysis, and communication. This section will be a bit of a laundry list, and I apologize in advance. I don't expect you to build all of these skills in a month; it will take longer to fully absorb the ideas and have the ability to apply them. Some of the topics may only come up for certain roles, and you'll know this when you see a job posting or have an interview.

Mathematics

Our treatment of calculus stopped short of power series and integral calculus. These topics will be important to learn. An understanding of integration will provide a foundation to better understand the continuous random variables of probability theory. The expected value of a continuous random variable is an integral, for example.

Many of the leading models in machine learning have deep underpinnings in probability, as they model sequences of events subject to randomness. Over time, you will need to learn advanced probability concepts and techniques. We studied gradient descent for fitting parameters in machine learning models. Some of the more complex models will use more specialized optimization techniques, such as stochastic gradient descent.

It will be necessary to go deeper in linear algebra to understand topics including norms, diagonalization, eigenvalues and eigenvectors, determinants, matrix factorization, and computation of the matrix inverse. The concepts behind these topics are important to understand, and implementation with software is equally important.

Statistics

We touched on statistical inference, but more specialized methods may be necessary for your work, including additional hypothesis tests, nonparametric testing, categorical data analysis, time series analysis, and Bayesian inference. It is important to know how to properly sample data and avoid bias. Some data scientists will find themselves designing experiments to determine if an effect is significant. These topics are covered in experimental design.

Computer Science

A data scientist should be comfortable with common data structures and algorithms, and should have an understanding of when to apply them. A good understanding of time complexity and space complexity will help you write better algorithms. The time complexity is the time it takes to run the algorithm as a function of the input length. The space complexity of an algorithm is the amount of memory consumed as a function of the input length. When comparing two algorithms, the one with lower time complexity and space complexity will be preferable.

We briefly covered databases and SQL. It will be very valuable to gain mastery of SQL. I should point out that large tech companies love to ask interview candidates about data structures, algorithms, and SQL.

It is essential for data scientists to be able to program well in Python, R, or something similar. Many data scientists do not regularly apply software development practices such as design patterns and unit testing, but these practices are highly encouraged. Writing efficient, reusable code takes practice and the study of great source material. There are many excellent open-source repositories available for such study. Moreover, for ML Engineering roles, strong software skills are a requirement.

Familiarity with a cloud-based system, such as AWS, Google Cloud, or Microsoft Azure, will be very valuable, as many organizations use the cloud for their computing and storage needs. The cloud providers evangelize their services and products, and their documentation can be very informative on topics including databases, analytics, machine learning, and hardware. As many of their services overlap, don't worry about learning each service offered by each provider.

Data Analysis

Similar to improving at coding, becoming better at data analysis requires a lot of practice and the study of good examples. It also requires knowledge and proper application of statistical concepts. Online communities such as Kaggle provide complete notebooks with code examples and datasets. There is a wide array of books filled with detailed code on practical data analysis and applied machine learning. As you read through the books, implement and test out the code.

Communication

Improving your communication skills can be done through regular practice and by asking for feedback. The practice might come through meetings, meetups, seminars, and other formal and informal events. Watch some skilled speakers and observe their habits and delivery. TED talks can be good sources of inspiration. Try to get experience presenting to both technical and non-technical audiences. Take note of the kinds of questions that each audience asks, so that you may better prepare in the future.

18.2 Advanced Technique: Regularization

One of the biggest risks faced by a machine learning model is overfitting. To mitigate this issue, we learned data splitting as a strategy. In this approach, the model performance is evaluated on a separately held out set of data. This provides an independent assessment of performance that is less likely to overfit nuances of the data.

Another powerful technique to prevent overfitting is *regularization*. Regularization takes the view that overfitting might be caused by model parameters that are too large. This may happen, for example, when the data sample is small and some of the observations are spurious. If you recall, OLS regression selects parameter estimates that minimize the sum of squared residuals. For a model with p predictors, the loss function looks like this:

$$SS_{res}(\boldsymbol{\beta}) = \sum_{i=1}^{n} (y_i - (\beta_0 + \beta_1 x_{i1} + \ldots + \beta_p x_{ip}))^2$$

To keep the parameter estimates small, regularization introduces a penalty term in this loss function. Its goal is to shrink the parameters toward zero. There are different penalty terms depending on the regularization technique used, and for illustration we will consider Lasso regression. It is worth noting an important feature of Lasso regression: if multiple predictors are highly correlated, only one of them will be included in the model. The remaining predictors will have coefficients of zero. If this effect is not desired, then a different penalty term will be warranted.

The penalty term in Lasso regression consists of two parts: the sum of the absolute value of the parameter estimates, which penalizes magnitudes, and a scalar multiple λ which controls the relative importance of the penalty term versus the sum of squared residuals term. The regularization term for p parameters looks like this:

$$\lambda \sum_{j=1}^{p} |\beta_j|, \quad \lambda \geq 0$$

The complete loss function for Lasso regression with p predictors combines the two terms as follows:

$$SS_{res}(\boldsymbol{\beta}) = \sum_{i=1}^{n} (y_i - (\beta_0 + \beta_1 x_{i1} + \ldots + \beta_p x_{ip}))^2 + \lambda \sum_{j=1}^{p} |\beta_j|, \quad \lambda \geq 0$$

The first term can be minimized by selecting parameter estimates which bring the predictions closer to the response values y_i. At the same time, the second term adds a positive quantity which increases with the sum of the magnitudes of the parameter estimates. Specifically, increasing any absolute coefficient $|\beta_j|$ will increase the penalty. Additionally, increasing λ will increase the value of the loss function. Ultimately, the optimal parameter estimates must strike a balance between these two terms.

Earlier, we encountered hyperparameters, which are important quantities in the configuration of a model. Additionally, their values are not known in advance. The λ term is an example of a hyperparameter, and a common way to estimate its value is through k-fold cross validation. Here is an outline of possible steps:

1. Provide a list of possible values for λ. We can start with powers of 10 since we are unsure of the right order of magnitude. This is the hyperparameter grid. Referring to this set as Λ, we might use $\Lambda = \{1 \times 10^{-4}, 1 \times 10^{-3}, \ldots, 1 \times 10^{3}, 1 \times 10^{4}\}$.

2. Decide on the number of folds k

3. Set aside a portion of the dataset for final evaluation (a test set)

4. For each possible value $\lambda_i \in \Lambda$, repeat these steps:

    For iterations $1, \ldots, k$, train the model on $k - 1$ folds and evaluate a metric such as MSE on the left-out fold. Each iteration should leave out a different fold.

    Average the MSE values to attain the final estimate MSE_i.

5. From the pairs (λ_i, MSE_i), select the λ_i with lowest value MSE_i. This yields the optimal value λ^*.

6. The final model is the version with regularization value λ^*

For a more accurate λ^*, a finer grid can be used in the first step after the MSEs are observed from the original course grid. Modern software packages support Lasso regression, as well as the popular variants Ridge regression and elastic net regularization. For more details on regularization, you might consult [4].

18.3 Advanced Machine Learning Models

Concurrent with strengthening the building blocks, I encourage you to explore and learn more classes of models. We studied regression models as they are prevalent, powerful, interpretable, and relatively simple in their architecture. For some data sets and use cases, other models will perform better. Two model classes which are extremely important are *tree-based models* and *artificial neural networks*. A brief overview will be provided next, and I recommend other sources for details, such as [4] and [49]. The latter provides an applied treatment with implementation in Python.

18.3.1 Tree-Based Models

Tree-based models are built up from the intuitive construct of a *decision tree*. Each tree forms a hierarchy, with a root placed at the top and branches emanating from the root and from each node in the tree. For a pair of nodes connected by a branch, there is a parent-child relationship where the parent is higher in the tree. The terminating nodes at the bottom of the tree are the leaves. These leaves serve as predictions. The nodes will represent useful predictors, while the branches will represent conditions on these predictors (e.g., if `age<20`, go left). A predictor is useful if it segments the observation target values into groups which increase in *purity*, where purity is defined by a metric such as *entropy*. For example, a 50/50 separation would have highest entropy (lowest purity), while a 100/0 separation would have lowest entropy (highest purity).

Figure 18.1 shows an example of a small decision tree using three variables: `age`, `visits`, and `score`. Beneath the leaves at the bottom of the tree is a row of boxes. Each box holds the distribution of the target values Y and N. For example, for observations with `age>=20` and `score<80`, there were 10 cases of Y and 10 cases of N. This path of the tree does not do a good job of separating the outcomes; we expect the probability of a correct prediction to be only 10/20. For a new observation, we can make a prediction by pushing it down the relevant path. For example, an observation where `age`=19 and `visits`=10 would first travel left and then right, receiving prediction N.

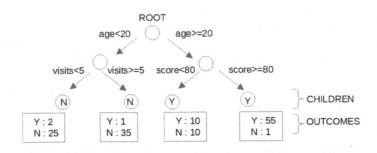

FIGURE 18.1: Sample Decision Tree

When fitting a decision tree to training data, important considerations include the predictors to use, the depth of the tree, and the cutpoints for each of the predictors (e.g., do we move left when `age<20` or when `age<21`?). A major shortcoming of the decision tree is that it can overfit to the training data. A useful technique for reducing overfitting is to build multiple decision trees and then aggregate the results into a prediction. Two popular and effective models following this approach are *random forests* (RF) and *gradient-boosted trees* (GBT). A variant of the GBT called the *XGBoost* is very efficient and often wins community data science competitions.

One distinction between RF and GBT is that the former builds a set of independent trees, while the latter builds a sequence of small trees meant to correct prediction errors. The "random" in random forests reflects the practice of drawing subsets of data and predictors when building each tree. After each tree is built, their predictions are equally weighted to produce an aggregated prediction. The GBT model, by contrast, will determine an optimal weighting of the small trees to produce an aggregated prediction. The technique of combining multiple trees to yield a prediction is called an *ensemble method*.

A major advantage of random forest models is that the construction of the independent trees can be divided among multiple computer cores to speed up processing. Since the GBT model builds trees sequentially as error refinements, the construction task cannot be done in parallel by multiple cores.

From their architecture, tree-based models do not assume a linear relationship between predictors and the target. This provides flexibility not present in linear regression and logistic regression models (recall that logistic regression models assume a linear relationship between the log odds and predictors). To be sure, there are many kinds of data where the linearity assumption is violated. For example, the relationship between a predictor and a target may exhibit this piecewise-linear behavior:

$$y = \begin{cases} 2x & x \leq c \\ 3x & x > c \end{cases}$$

In this case, the slope gets steeper when x exceeds c. Linear regression would not handle this relationship properly, but a tree-based model could learn the non-linear pattern from the data and use branching as needed.

Regression models require the user to determine the best set of predictors.[1] Tree-based models use an algorithm to automatically determine useful predictors, placing the most important predictors higher in the tree. Predictors that aren't sufficiently helpful for increasing explanatory power are dropped from the model. For a large set of potential predictors, this can save a lot of development time.

Regression models require imputation of missing values. For some implementations of tree-based models, missing values are treated as a separate level for each predictor. In doing this, imputation is not necessary in the tree-based model.

A challenge introduced by tree-based models is that interpretation is less clear when compared to linear and logistic regression models. This is primarily because tree-based models combine output from multiple trees when making a prediction. There are specialized methods for measuring the importance of predictors in these models, but they are still more challenging to understand than the parameter estimates in a regression model.

18.3.2 Artificial Neural Networks

Artificial neural networks, simply called neural networks when there is no risk of confusion with biological neural networks, were first developed in the 1950s and have seen a resurgence in the last several decades. Many of today's state-of-the-art models are based on customized neural networks. We will learn about some of their use cases later and see an overview of their architecture here.

The building blocks of neural networks are nodes, weights, and biases. The nodes are arranged in layers, beginning with an *input layer*, followed by one or more *hidden layers*, and finally an *output layer*. Figure 18.2 shows a small illustration of a neural network with one hidden layer. The input layer takes the predictors as input, and there is one node per predictor. The number of hidden layers and the number of nodes per hidden layer are hyperparameters to be decided through tuning. Each hidden layer may have a different number of nodes. Neural networks with more than one hidden layer are called *deep neural networks* (DNNs). The subfield of machine learning that uses DNNs is called *deep learning*. The number of nodes in the output layer will match the task. If the purpose of the model is to predict a binary output, for example,

---

[1]Including a regularization term in the model can allow for automated predictor selection.

then one node is sufficient to provide the probability of the positive class p; the probability of the negative class will then be $1 - p$.

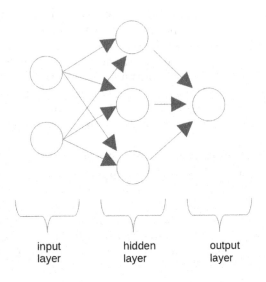

FIGURE 18.2: Sample Neural Network

The calculation strategy used by neural networks is to apply a set of weights to the values in a layer of nodes and add a bias term b; the bias term plays the same role as the intercept term from regression. The result is an expression like this: $b_i + w_1 x_1 + \ldots + w_p x_p$. This expression may remind you of the linear combinations of predictors from regression. The difference is purely notational, with w_i playing the role of β_i. The next steps are very different from regression, however: a non-linear *activation function* is applied to the expression. There are several commonly used activation functions, and their purpose is to give the model flexibility to capture non-linear relationships between inputs and the target. After the inputs are weighted, adjusted by a bias term, and passed through the activation function, the intermediate quantities may pass through these same steps one or more times. This is a composition of functions, and different sets of weights and biases are used in each layer of the network. To organize the calculations and the data, the weights are stored in a matrix **W** and the biases are stored in a vector **b**. Given an increasing number of parameters, useful activation functions, and a host of other features, neural networks can fit extremely rich, complex patterns in data.

The mathematics necessary to study neural networks is mostly things you already know. The *feedforward* phase of pushing inputs through the model and arriving at predictions uses arithmetic operators, matrix algebra, and the compositions of functions. The more intricate part is optimizing the weights and biases so that the predictions are close to the target values. The optimization uses a loss function and a variant of gradient descent (most commonly). To decide how the parameters should change, partial derivatives of the loss function are computed with respect to each parameter. Much of the complexity arises from the layers in the architecture: there is a sequence of derivatives that must be computed on composed functions. You may recall that to calculate the derivative of composed functions, the chain rule can be used. The chain rule, combined with matrix algebra, is essential in optimizing the parameters of a neural network. The breakthrough method for enabling these calculations is called *backpropagation*. When you study neural networks, I highly recommend that you spend time understanding how backpropagation works, as this is fundamental to fitting neural networks.

The model just described is a fully connected neural network, as each node in a layer is connected to each node in the next layer. The leading models in areas such as computer vision use neural networks that are not fully connected. Specifically, the connections may be placed based on nearby locations or time points. These architectures significantly reduce the number of connections, which reduces storage and computation. For example, a computer vision model used to detect objects may not need pixel information from one image corner to predict the contents of a different image corner. These models include other adaptations as well to improve performance and accommodate inputs of different shapes. However, you have already learned many of the required concepts to dive deeper into neural networks.

18.4   Additional Languages

Python and SQL will take you far. I encourage you to gain mastery of these two languages. Another popular language for data science is R, and it has roots in statistical computing. R offers a wide array of advanced models and beautiful graphics. You might find that you need to learn R or another language for a project or a job. It can be helpful to your career to stay flexible and pick up new tools as needed. After digesting the important concepts of a programming language, it is easier to learn additional languages.

If you need to compute with massive datasets exceeding the machine's memory, Python and R will not be sufficient. The most popular programming language for big data computing is Apache Spark. The benefits to Spark are numerous: it is free, open source, it offers APIs in several languages including R and Python, it easily distributes data and code to a cluster of computers,[2] and it supports a range of tasks. In particular, Spark has modules that

---

[2]In the distributed computing paradigm, the data is divided and sent to each worker machine. The completed work can be collected back to a single machine called the *driver*.

support analytics, SQL, machine learning, graph processing, and stream processing. It can ingest a variety of formats and send results downstream to applications and dashboards. Spark is so useful that I developed big data courses around it for the University of Virginia and the University of California, Santa Barbara. You can find the Big Data Systems course at UVA here: `https://github.com/UVADS/ds5110`. An excellent book for learning Spark is [50].

18.5   Resources

Certification and Degree Programs

In addition to the textbooks mentioned throughout this book, there are several other places to learn data science. There is a wide array of online courses from vendors including Coursera, Udemy, Udacity, and Codeacademy. There are several free courses on YouTube. A course that I found particularly valuable early in my data science career is the Stanford CS229 Machine Learning course taught by Andrew Ng. This course offers a detailed, mathematical treatment of many essential models and techniques.

Several companies and universities offer bootcamps, certifications, and degree programs. There are several online master's programs, such as the program where I teach at the UVA School of Data Science. When deciding to make a large investment of resources, it is best to understand costs, curriculum content, support, networking opportunities, and exit opportunities. It can be helpful to review testimonials and placement statistics from the various programs.

Cloud Computing

There are many no-cost and low-cost resources for computing in the cloud. AWS offers a free tier to try their services:

AWS Free Tier: `https://aws.amazon.com/free`.

Their end-to-end machine learning service, Amazon SageMaker, includes a free notebook environment which runs in a browser:

`https://studiolab.sagemaker.aws/`

SageMaker Studio Lab supports computing with CPUs and GPUs.

Google Colab is another free option for computing in your browser. Users can access CPUs and GPUs. Notebooks are saved in Google Drive for easy sharing. Colab can be accessed here:

`https://colab.research.google.com/`

Rounding out the three largest cloud providers, Microsoft also offers free access to some of their services. Some services are always free, while others are free for the first 12 months. You can browse the services here:

`https://azure.microsoft.com/en-us/pricing/free-services/`

For each of these services, it is best to confirm the pricing structure in advance.

Communities

Kaggle is an online community of data science and machine learning students and practitioners. It is free to join and use the resources. Kaggle hosts competitions, projects, datasets, and notebooks. This can be a great way to develop your skills. Many of my students find interesting data sets on Kaggle and use them for group projects.

There are many other sites for online data science competitions (which I haven't tried), such as DrivenData, Devpost, and Numerai.

18.6   Applications

The breadth of data science and machine learning applications is staggering. If you name a problem, there is likely a model for it, or the opportunity for a model. Below, I provide a few broad, exciting areas with high activity. Deep learning is the leading approach for each of them.

Recommendation

We have encountered elements of recommender systems throughout the book. The first thought that comes to mind on the topic might be recommendations from Netflix and Amazon, but this is just the beginning. In the broadest sense, recommender systems are useful anywhere that there are products or services which may be of interest. As the catalog of products grows, it becomes increasingly difficult to find relevant material. Education technology companies want to make it easier for teachers to find helpful lesson plans. Government entities want to help citizens find relevant documents. A nice book for going deeper on recommendation is [51].

Computer Vision

Computer vision is being used to detect objects of interest in images and videos. The models can provide a list of found items, their locations, and attributes about them. For example, a model might indicate that an adult and child are in a photo, that the adult is wearing glasses, and that the child has her eyes closed. A model might flag an unexpected person taking an exam as an indicator of potential cheating. A traffic signal might count the number of vehicles at a busy intersection to help reduce traffic congestion and increase safety. The potential applications of this technology are massive. The extracted information can be stored with the underlying data as metadata and tags. The enriched dataset can be used to improve search and recommendation.

Natural Language Processing and Understanding

One of the earliest language models, called the *bag of words* model, uses the count of each word in a document to provide information about the document. An article containing several mentions of the word "football" might be classified as an article about sports. For a more robust understanding of the text, the context of each word needs to be incorporated. One of the challenges for a language model is understanding the relationship between far-apart words or phrases in the text. For example, consider this text:

"Jacqui loved visiting *Barcelona* in the summertime. She enjoyed the way that flamenco guitarists would play music in the streets. Yesterday, when I asked what she would like to do next summer, she mentioned how much she would like to go back *there*."

If we asked a bag of words model to identify the meaning of *there*, it would not be able to make the connection to Barcelona.

More advanced models build an understanding of context with objects called *embeddings*. Several decades of work have led to state-of-the-art *transformer* models consisting of stacks of processing modules. The leading models currently use hundreds of billions of parameters and can accomplish several language tasks including language translation, text summarization, and detection of entities of interest.

Speech Recognition

Speech-to-text recognition services support the recognition and translation of spoken language into text. Personal assistants such as Siri and Alexa recognize users' voices, process the speech, and provide personalized responses. The technology behind speech recognition is deep learning models that process sequences of text. Challenges include the understanding of diverse accents and dialects.

Protein Folding

A string of dozens to hundreds of amino acids determines the complex 3D shape of proteins in the human body. The shape determines how the proteins function, and knowledge of this shape can help researchers devise effective medications [52]. The number of amino acids and their possible interactions posed a massive combinatorial problem that has vexed scientists since the 1960s. In 2021, a model named AlphaFold was delivered by DeepMind. It uses a deep learning model trained on a massive repository of data to predict the 3D shape of proteins. In 2022, the team released protein structure predictions for nearly all catalogued proteins known to science.

Generative AI

The transformer models mentioned earlier are behind breakthrough deep learning models that can take a prompt from a user and generate entirely new content including text, code, and images. The Stable Diffusion model can create a detailed image from a text description. The GPT-4 model can accept image and text inputs and produce text outputs. ChatGPT is a high-powered chatbot that took the world by storm in 2022 with its fluent responses to user prompts. The applications for generative AI are vast and include drafting documents, summarizing content, developing code, and providing educational assistance.

Like any groundbreaking technology, Generative AI offers a powerful tool for accelerating work. At the same time, the current models have shortcomings that need to be addressed. The training data can contain biased or toxic content which can be propagated by a model. If a model does not have sufficient information to answer a question, it can *hallucinate*, or craft incorrect responses. Human reviewers, including data scientists, can help to improve these models to make them safer and more useful.

18.7 Final Thoughts

I hope that this chapter provided some good ideas of where to go next, and that it stoked excitement about where the field is headed. Finally, I sincerely hope that you enjoyed reading this book as much as I enjoyed writing it. Please feel free to reach out to me on LinkedIn, including a note that you read the book.

https://www.linkedin.com/in/adam-tashman-93a82722

You can learn about my company, PredictioNN, and browse data science resources at this URL:

```
http://prediction-n.com/
```

The PredictioNN website includes links to the data science course, supporting material for this book, and other helpful resources for learning data science.

Bibliography

[1] How much data do we create every day? The mind-blowing stats everyone should read. https://tinyurl.com/2s35wxd4. Accessed: 2022-11-28.

[2] Regression toward the mean. https://en.wikipedia.org/wiki/Regression_toward_the_mean. Accessed: 2022-11-28.

[3] Machine learning lens. https://tinyurl.com/2p8wb69h. Accessed: 2022-11-28.

[4] Trevor Hastie, Robert Tibshirani, Jerome H Friedman, and Jerome H Friedman. *The elements of statistical learning: Data mining, inference, and prediction*, volume 2. Springer, 2009.

[5] Kerry Patterson, Joseph Grenny, Ron McMillan, and Al Switzler. *Crucial conversations tools for talking when stakes are high*. McGraw-Hill Education, 2012.

[6] Simon Sinek. *Start with why: How great leaders inspire everyone to take action*. Penguin, 2011.

[7] Dena M Bravata, Sharon A Watts, Autumn L Keefer, Divya K Madhusudhan, Katie T Taylor, Dani M Clark, Ross S Nelson, Kevin O Cokley, and Heather K Hagg. *Prevalence, predictors, and treatment of impostor syndrome: A systematic review*, volume 35. Springer, 2020.

[8] What is an abundance mindset? https://www.tonyrobbins.com/mind-meaning/adopt-abundance-mindset/. Accessed: 2023-04-13.

[9] Jocko Willink and Leif Babin. *Extreme ownership: How US Navy SEALs lead and win*. St. Martin's Press, 2017.

[10] Jocko Willink. *Leadership strategy and tactics: Field manual*. Pan Macmillan, 2020.

[11] Tensorflow model garden. https://github.com/tensorflow/models. Accessed: 2023-02-01.

[12] States of digital data. https://tinyurl.com/3kr43fdf. Accessed: 2023-01-10.

[13] Kevin Mitnick. *The art of invisibility: The world's most famous hacker teaches you how to be safe in the age of big brother and big data*. Little, Brown, 2017.

[14] FAQ: How do i know if my sources are credible/reliable? `https://guides.lib.uw.edu/research/faq/reliable`. Accessed: 2022-12-15.

[15] What is a REST API? `https://www.redhat.com/en/topics/api/what-is-a-rest-api`. Accessed: 2023-07-04.

[16] Wikipedia 1.4.0. `https://pypi.org/project/wikipedia/`. Accessed: 2023-07-04.

[17] Daniel Jurafsky and James H Martin. *Speech and Language Processing: An Introduction to Natural Language Processing, Computational Linguistics, and Speech Recognition*. Pearson, 2014.

[18] What is processor speed and why does it matter? `https://tinyurl.com/mwaj3ufe`. Accessed: 2022-12-20.

[19] CPU vs GPU: What's the difference? `https://www.intel.com/content/www/us/en/products/docs/processors/cpu-vs-gpu.html`. Accessed: 2022-12-20.

[20] William Shotts. *The Linux command line: A complete introduction*. No Starch Press, 2019.

[21] Scott Chacon and Ben Straub. *Pro Git*. Springer Nature, 2014.

[22] An intro to Git and GitHub for beginners. `https://product.hubspot.com/blog/git-and-github-tutorial-for-beginners`. Accessed: 2022-12-20.

[23] Anna V Vitkalova, Limin Feng, Alexander N Rybin, Brian D Gerber, Dale G Miquelle, Tianming Wang, Haitao Yang, Elena I Shevtsova, Vladimir V Aramilev, and Jianping Ge. *Transboundary cooperation improves endangered species monitoring and conservation actions: A case study of the global population of Amur leopards*, volume 11. Wiley Online Library, 2018.

[24] Wes McKinney. *Python for data analysis: Data wrangling with Pandas, NumPy, and IPython*. O'Reilly Media, Inc., 2012.

[25] Rod Stephens. *Beginning database design solutions*. John Wiley & Sons, 2009.

[26] Richard Courant. *Differential and Integral Calculus, Volume 1*. Ishi Press, 2010.

[27] Richard Courant. *Differential and Integral Calculus, Volume 2*. John Wiley & Sons, 2011.

[28] Joseph K Blitzstein and Jessica Hwang. *Introduction to probability*. CRC Press, 2019.

[29] Stephen Friedberg, Arnold Insel, and Lawrence Spence. *Linear Algebra*, volume 4. Prentice Hall, 2002.

[30] George W Snedecor and William G Cochran. *Statistical methods*. Iowa State University Press, 1989.

[31] Richard Wesley Hamming. *Digital filters*. Courier Corporation, 1998.

[32] An introduction to seaborn. `https://seaborn.pydata.org/tutorial/introduction.html`. Accessed: 2023-02-01.

[33] matplotlib getting started. `https://matplotlib.org/stable/users/getting_started`. Accessed: 2023-02-01.

[34] John F Helliwell, R Layard, and J Sachs. *World Happiness Report 2018*. Sustainable Development Solutions Network New York, 2018.

[35] Edward R Tufte. *The visual display of quantitative information*, volume 2. Graphics Press, 2001.

[36] Olivier Chapelle, Bernhard Scholkopf, and Alexander Zien. Semi-supervised learning. *Cambridge, Massachusetts: The MIT Press View Article*, 2, 2006.

[37] Richard S Sutton and Andrew G Barto. *Reinforcement learning, second edition: An introduction*. MIT press, 2018.

[38] Vladimir N Vapnik. *Statistical learning theory*. Wiley, 1998.

[39] Norman R Draper and Harry Smith. *Applied regression analysis*, volume 326. John Wiley & Sons, 1998.

[40] Douglas C Montgomery, Elizabeth A Peck, and G Geoffrey Vining. *Introduction to linear regression analysis*. John Wiley & Sons, 2021.

[41] R Kelley Pace and Ronald Barry. *Sparse Spatial Autoregressions*, volume 33. EconPapers, 1997.

[42] Kaare Brandt Petersen, Michael Syskind Pedersen, et al. The matrix cookbook. *Technical University of Denmark*, 7(15):510, 2008.

[43] Matjaz Zwitter and Milan Soklic. Breast cancer. UCI Machine Learning Repository, 1988. DOI: https://doi.org/10.24432/C51P4M.

[44] CI/CD concepts. `https://docs.gitlab.com/ee/ci/introduction/`. Accessed: 2023-07-19.

[45] Docker for data science: An introduction. `https://www.datacamp.com/ tutorial/docker-for-data-science-introduction`. Accessed: 2023-07-19.

[46] Cathy O'Neil. *Weapons of math destruction: How big data increases inequality and threatens democracy.* Crown, 2017.

[47] Groundhog day forecasts and climate history. `https://www.ncei.noaa. gov/news/groundhog-day-forecasts-and-climate-history`. Accessed: 2023-02-01.

[48] Amazon Sagemaker Clarify. `https://aws.amazon.com/sagemaker/ clarify/`. Accessed: 2023-04-19.

[49] Sebastian Raschka. *Python machine learning.* Packt publishing ltd, 2015.

[50] Jules S Damji, Brooke Wenig, Tathagata Das, and Denny Lee. *Learning Spark.* O'Reilly Media, Inc., 2020.

[51] Kim Falk. *Practical recommender systems.* Simon and Schuster, 2019.

[52] The game has changed. AI triumphs at solving protein structures. `https://tinyurl.com/99cwkyr4`. Accessed: 2023-04-05.

Index

Printed in the United States
by Baker & Taylor Publisher Services